Towards the Local Government's Integrated Accountability Framework

Habib Muhammad Shahib

Towards the Local Government's Integrated Accountability Framework

A Critical Lesson from Socio-Environmental Issues in Indonesia

Springer

Habib Muhammad Shahib
Faculty of Economics and Social Science
Universitas Fajar
Makassar, Indonesia

ISBN 978-981-33-6616-9 ISBN 978-981-33-6617-6 (eBook)
https://doi.org/10.1007/978-981-33-6617-6

This Springer imprint is published by the registered company Springer Nature Singapore Pte Ltd.
The registered company address is: 152 Beach Road, #21-01/04 Gateway East, Singapore 189721, Singapore

To:

my Parents,
my wife, Halirah,
my son, Sam, and
my daughter, Nasywa.

Preface

By thanking Allah, the almighty God, for His abundance and mercy, also the great inspiration from the Prophet Muhammad (peace be upon him), I can present a book entitled:

Towards the Local Government's Integrated Accountability Framework: A Critical Lesson from Socio-Environmental Issues in Indonesia.

This book topic starts with my concern about the condition of the world in general, and Indonesia in particular, which shows the growing phenomenon and the broad impact of socio-environmental conflicts in the grassroots communities (farmers, fishermen and urban poor). It is because of the government's development orientation, which tends to be oriented towards economic growth, and denies sustainable development. In particular, the sustainable development, which is in line with the aspirations of small/poor communities. This situation is seen in the daily life of the public. The development of the extractive industries, from mining, plantations, agriculture, oil-gas to infrastructure, tends to ignite socio-environmental conflicts with local communities. Also, these extractive activities are generally ending by the displacement of local communities in their living spaces and damaging the environmental durability in the community.

The big question then arose in my mind. In the middle of the discussion on the discourse of sustainable development, especially in accounting studies, is there any way to give a respected place to the aspirations of small communities affected by economic development in the discourse of sustainable development. Hence, their voices are not only heard as protests but can be the basis of reference to government development policies.

After being immersed in reading a variety of literature and accounting studies, I had two main clues to introduce and, I believed, these ideas could change the direction of government development towards more empowerment and support to the environmental resilience of small communities. First, the concept of government accountability, this concept is based philosophically on the assumption of an integrated concept from upstream to downstream, starting from the planning-policies-practices of government accountability reports (AccountAbility, 2008;

Dwivedi, 1989; KNKG, 2008; OECD, 2018). Based on this concept, the public could push the government to be more responsive to a problem and to adapt its policies in line with the aspirations of the people. With a chain model of the government-policy-practice-accountability reporting, the change movement can start in different directions. From top to down, from bottom to up, even, as a whole, these can be choices to change the direction of government development.

The second clue I found is the concept of emancipatory socio-environmental accounting developed by Gray (2002), Tinker and Gray (2003), Saravanamuthu (2009), and Saravanamuthu and Lehman (2013). This critical concept encourages us to focus on the socio-local context in the development of alternative models of accountability, which give more space to people's aspirations. One of the goals of this concept is how people [have ideas to] control their environment. Thus, there will be a unique model of socio-environmental accountability in each region (Saravanamuthu & Lehman, 2013). Interestingly, the concept of emancipatory socio-environmental accounting is also very encouraging to use alternative methodologies in the development of socio-environmental accounting/accountability, particularly, methodologies that have an ontology-epistemology-axiology-locality coherence. Consequently, the choice then falls into the critical thought of Tan Malaka, philosopher, and Indonesian freedom hero, whose thoughts that are very consistent with the construction of government systems and the socio-political-economic context of Indonesia.

With these two main guidelines, I then conducted this study to develop the accountability framework of local governments in Indonesia. Therefore, it can provide more space to accommodate the aspirations of people affected by socio-environmental problems due to the local governance and development policies' power in the current era of regional government autonomy is becoming powerful from time to time. Also, it is because of Indonesia's unique socio-political context, explicitly the civil society movement, which is generally led by socio-environmental non-governmental organizations (NGOs) to advocate the issue of government development versus socio-environmental issues of the community. I then focused on exploring the counter-discourses advocated by socio-environmental NGOs together with the affected communities in Indonesia. By hoping, it can further develop a framework of local government accountability that can be more empowering the socio-environmental issues in Indonesia and integrating these issues into the government's accountability reports.

Finally, in this book, the main points addressed include: (1) the background and motivation of this book, (2) the review of the literature which shows the revolution of accountability model from time to time. Particularly, it is related to the model and discourse of New Public Management (NPM), and sustainable development in government. Next, this book displays (3) a unique research methodology that uses the critical thought of Tan Malaka, as Indonesia's national hero, to develop the framework of integrated accountability for local government in Indonesia. Also, it shows (4) a description of research results that illustrates the phenomenon of local government accountability in Indonesia, which is tasted of NPM, and this condition is leading to marginalization of socio-environmental issues in the government's accountability system. At the end part, this book shows (5) the discussion of the synthesis of research results (6) which leads to the new development of an integrated local government accountability framework from the social, environmental and economic aspects with its primary support: substantial participation and transparency, optimization of data infrastructure and technology and adding the government accountability reports (the socio-environmental impact of developmental projects and the socio-environmental sustainability reports). Last, this book is closed by the conclusions and suggestions for future research.

The completion of this study would also have been difficult without generous supports from the following persons and parties. Bung Muhammad Al-Amin, as the Director of WALHI-South Sulawesi (2018–2022), Thank you very much for your help in giving me access to relevant information and informant related to the socio-environmental issues in south Sulawesi, and Indonesia in general. *Daeng* Sitaba and all friends in FORMASI NEGARA (Forum Masyarakat Nelayan Galesong Raya) thank you very much for showing me the other reality of socio-environmental issues behind the economic development in Indonesia, specifically in south Sulawesi. Finally, thanks to LPDP (Indonesia Endowment Fund for Education) for its funding support. Also, thanks to research colleagues, for the great support during this study.

Last but not least, I recognize that this book is not perfect. Therefore, I hope for suggestions and constructive criticism from all readers and scholars. Finally, I hope that this book can bring advantages to the Indonesian people and to the development of accounting studies, specifically related to the local government's accountability discourse in Indonesia, and the world in general.

Regards,

Makassar, Indonesia
April 2020

Habib Muhammad Shahib

References

AccountAbility. (2008). *AA1000 Accountability Principles Standard 2008*. Retrieved from www.accountability21.net.

Dwivedi, O. P. (1989). Public Service Responsibility and Accountability. In J. G. Jabbra & O. P. Dwivedi (Eds.), *Public service accountability: A comparative perspective*. Connecticut: Kumarian Press.

Gray, R. (2001). Thirty years of social accounting, reporting and auditing: what (if anything) have we learnt? *Business Ethics: A European Review*, *10*(1), 9–15. https://doi.org/10.1111/1467-8608.00207.

KNKG. (2008). *Pedoman Umum Good Public Governance Indonesia*. Jakarta: Komite Nasional Kebijakan Governance (KNKG).

OECD. (2018). *Accountability (in Management Theory)*. Retrieved May 11, 2018, from Glossary of Statistical Term-OECD website: https://stats.oecd.org/glossary/detail.asp?ID=4757.

Saravanamuthu, K. (2009). A journey of socialising the risks associated with global warming: a Gandhian insight into Schumacher's total accounting and accountability. In *Extending Schumacher's Concept of Total Accounting and Accountability into the 21st Century* (pp. 121–189). https://doi.org/10.1108/S1041-7060(2009)0000014008.

Saravanamuthu, K., & Lehman, C. (2013). Enhancing stakeholder interaction through environmental risk accounts. *Critical Perspectives on Accounting*, *24*(6), 410–437. https://doi.org/10.1016/j.cpa.2013.02.002.

Tinker, A., & Gray, R. (2003). Beyond a critique of pure reason: from policy to politics and praxis in environmental and social research. *Accounting, Auditing, and Accountability Journal*, *16*(5), 727–761.

Contents

List of Figures

List of Tables

List of Pictures

Chapter 1
Introduction

Once the last tree is cut, the last fish is trapped,
the last river is poisoned;
Once to inhale the air is shocking,
You will recognize, too late,
That prosperity is not in your bank accounts,
and you cannot consume money!
(Native America's Quote) (This quote is very common used in GREENPEACE's campaigns and other environmental activist's campaigns (Greenpeace, 2017). This quote firstly known in a chapter essay that entitled "conversations with north America Indians" in a collection of essays that titled "who is the chairman of this meeting?: A collection of Essays" (Osborne, 1972)).

1.1 Research Background

In the end of 1990s, after the 98-reformation movement, Indonesia implemented a new policy called *otonomi daerah* (regional autonomy). This marks a new political era (Hadiz, 2004; Rasyid, 2003; Sanusi, Butler, & Rola-rubzen, 2013). This policy has given a broader space to local governments, provinces-districts-cities, to manage their area by themselves (Rasyid, 2003). Therefore, local governments become more authoritative in giving a direct policy in their field, especially in economic policy (Sanusi et al., 2013).

Consequently, local economic policy becomes more dominant. Thus, the local governments can easily invite and accept the (local and international) investor to invest in their area (World-Bank, 2012), notably, in the natural resources sector such as forestry, agriculture, plantation, marine, and mining, which always become leading economic sources and employment provider (WALHI, 2015, 2016; World-Bank, 2012). As the 2016 data indicate, natural resources sectors become the first provider of the non-tax revenue (36.73%), followed by agricultural sector (13.2%) and mining sector (7.5%). They are the three main contributors of Indonesia's GDP (BI, 2016; KataData, 2017).

H. M. Shahib, *Towards the Local Government's Integrated Accountability Framework*,
https://doi.org/10.1007/978-981-33-6617-6_1

Table 1.1 Socioeconomic conflicts and environmental damages related to investment activities in the natural resources sector in Indonesia

	The impact extent	In conflict with/Caused by
Agrarian conflicts in 2016 (KPA, 2017)	1,265,027.39 Ha (throughout Indonesia)	In conflict with Industries (51.78%), Government (22.44%), inter-local people (14.44%), Police and Military (5.78%), and others (5.56%)
Forest fires in 2015 (WALHI, 2016)	2,700,000 Ha and 38 people dead (in 17 provinces)	Caused by the business actors who did not obey the environmental requirements and the lack of government's environmental policy
Floods and Landslides (WALHI, 2016)	580 times and 226 people dead (throughout Indonesia)	Supposedly, caused by the business actors who did not really obey the environmental requirements and the lack of government's environmental policy

In contrast, there is an opinion that the growing number of commerce in the natural resource field contributes to the growth of socio-environmental negative effects in the local regions. Mostly, the natural resource industry results in socioeconomic conflicts related to the land, natural resources, and environmental damage, as extensively advocated and reported by local Non-Governmental Organizations (NGOs) (LBH-Jakarta, 2015; LBH-Makassar, 2016; WALHI, 2015, 2016), as shown in Table 1.1.

Due to the above issues, many socio-environmental activists protest against the government, especially to the local governments' socio-environmental policies and activities, although, the Indonesia government has been trying to be extra transparent and accountable to the people, since the 1998 post-political and economic crisis. The activists consider that the practical system of socio-environmental management and accountability of the local government are inappropriate. These result in conflicts related to natural resources between indigenous people and firms, local people banished from their home and cultivated fields, forest fires, local water-food scarcity and ecosystems destruction: these problems are difficult to be controlled (KPA, 2017; LBH-Jakarta, 2015; LBH-Makassar, 2016; Republika Online, 2015; Timur, 2015; WALHI, 2015, 2016).

To achieve the accountability and transparency, many government's accountability instruments are used such as financial statement,[1] supervisory reports related to financial accountability in central and local governments,[2] and performance accountability

[1]Indonesia Government Regulation Number 71/2010 regarding The Government Accounting Standard (accrual basis).

[2]Indonesia Government Regulation Number 60/2008 regarding Internal government monitoring system.

reports of government agencies.[3] Even, each year the KEMENPAN-RB[4] together with government internal auditors assesses the accountability reports of local governments and give rewards to the best accountability reports[5] (Kemenpan-RB, 2018a, 2018b). However, the accountability instruments are still regarded as being based only on monetary and economic measurements (Diefenbach, 2009; Djamhuri, 2009; Harun, An, & Kahar, 2013; Lapsley, 2009; Rajiani, 2011) and lacking functions in terms of accountability in social-environmental issues in the local context in Indonesia, precisely by the socio-environmental campaigners' views (KPA, 2017; LBH-Jakarta, 2015; LBH-Makassar, 2016; WALHI, 2015, 2016; Watchdoc, 2018).

Therefore, the Indonesian socio-environmental activists' viewpoints become an interesting aspect to be further discussed. The viewpoints can be used as a basis for examining the governments' accountability tools related to socio-environmental accountability as there is a possibility that activists' voices tend to be ignored by the government due to the domination of economic and monetary logic or the New Public Management paradigm in the government's development policy making (Diefenbach, 2009; KPA, 2017; LBH-Jakarta, 2015; LBH-Makassar, 2016; Rajiani, 2011; Sari, Djamhuri, & Kamayanti, 2014; WALHI, 2015, 2016; Watchdoc, 2018). Thus, the socio-environmental issues do not become the primary focus of the local governments in their accountability instruments.

According to the above reasons, the concern for conducting a further study related to the socio-environmental activists' voices also aligns with the concern of accounting scholars in the last three decades. The accounting scholars in most socio-environmental accounting literature agree that accounting studies should be involved in the efforts to stimulate and support the social-economic-environmental sustainability. It is because the accounting discourse can present a useful accounting instrument for maintaining the relationship between the government, environmental, and social structure of the society in the economic-operating area (Gray, 2001; Gunderson, Holling, & Editors, 1995; MacLullich, 2004; Parker, 2005; Saravanamuthu, 2004, 2009; Saravanamuthu, Brooke, & Gaffikin, 2013; Saravanamuthu & Lehman, 2013; Stiglitz, Sen, & Fitossi, 2009; Sukoharsono, 2005, 2007).

However, the development of socio-environmental accounting which pays attention to the instrument of socio-environmental accounting, particularly focusing on the socio-environmental accountability model in the local government is still rare (Qian, Burritt, & Monroe, 2011; Williams, Wilmshurst, & Clift, 2011). It tends to be based on voluntary work or follow the worldwide trend (Niemann & Hoppe, 2018). Especially,

---

[3]The Regulation of the Minister of State-Apparatus Employment (*Permenpan*) Number 29/2010 regarding The Guidelines of Performance Determination Compilation; and Performance Accountability report in the Government Agencies (the latest revision of *Permenpan* Number 53/2014).

[4]*KEMENPAN-RB (Kementerian Pendayagunaan Aparatur Negara dan Reformasi Birokrasi) or* (Ministry of Administrative and Bureaucracy Reform).

[5]For example, in 2018, the province governments of East Java, DIY, West Java and South Sumatra, and the regency/city governments, namely Bandung and Banyuwangi got grade A and Regional Incentive Fund (RIF) for their accountability report (*LAKIP-Laporan Akuntabilitas Kinerja Instansi Pemerintah*) (Kemenpan-RB, 2018b).

studies which focus on the socio-environmental accountability forms in local government agencies which are directly involved in managing the social and environmental issues (Ball, 2004, 2005, 2007; Qian et al., 2011; Shahib & Risky, 2017; Williams et al., 2011). In the Indonesia context, the socio-environmental accounting discourse tends to focus on the disclosure of corporate social responsibility (CSR) of private firms that listed on the Indonesia Stock Exchange (BEI) (Shahib & Irwandi, 2016; Sukoharsono, 2007), although, the socio-environmental discourse has existed and developed since the 1980s (Deegan, 2002; Gray, 2001; Saravanamuthu & Lehman, 2013). There are also movement from several local governments in other countries that have tried to make an integrated or sustainability report[6](Niemann & Hoppe, 2018).

Saravanamuthu and Lehman (2013) argue that further studies on the basic theory and measurement development of the socio-environmental accountability that fits the "present" and "locality" circumstances should be conducted. This is supported by Burritt (2004), Deegan (2002), Gray (2001), Parker (2005), and Sukoharsono (2005, 2007) who emphasize that the socio-environmental accounting studies should pay attention to the practices and issues in reality. Thus, the opportunities to do a direct study associated with the practical accounting and socio-environmental accountability model are still widely open, particularly, in the local socio-environmental and government settings (Niemann & Hoppe, 2018; Qian et al., 2011; Williams et al., 2011), especially, in the Indonesian context where socio-environmental accountability schemes are still greatly questioned by the local socio-environmental Non-Government Organizations (NGOs), the society and academicians (KPA, 2017; LBH-Jakarta, 2015; LBH-Makassar, 2016; Sukoharsono, 2007; WALHI, 2015, 2016).

In line with the previous explanations, it is believed that the accounting discourse can produce a useful accountability mechanism for maintaining the relationship between the government, environment, and social structure of the society in the economic-operating area (Gray, 2001; Gunderson et al., 1995; MacLullich, 2004; Parker, 2005; Saravanamuthu, 2004, 2009; Saravanamuthu & Lehman, 2013; Saravanamuthu et al., 2013; Stiglitz et al., 2009; Sukoharsono, 2005, 2007). Hence, ontologically, the basic idea of accountability tool can be classified as the idealism or the constructed reality (Burrell & Morgan, 1979; Creswell, 2013). Based on this reason, this research uses a natural research setting. Therefore, an ideographic methodology is employed in this research (Creswell, 2013). Moreover, as the focus of this research is to deliver the socio-environmental accountability based on the activists' views, which is tend to be ignored by the government (KPA, 2017; LBH-Jakarta, 2015; LBH-Makassar, 2016; WALHI, 2015, 2016), then the critical paradigm will be used in this research (Kamayanti, 2016, 2017).

[6]Amsterdam, Basel, Dublin, Freiburg, Nuremberg and Zurich are the Several world cities government that have implemented the sustainability report (Niemann & Hoppe, 2018).

1.2 Research Motivations

Based on the issue and research gap explained above, it is interesting to conduct research related to the socio-environmental accountability context in the local government setting in Indonesia, in particular, with the intention of using the socio-environmental activists' aspiration as the basis for evaluating and developing the socio-environmental accountability policies. In Indonesia, there are massive movements from the socio-environmental activists and NGOs in criticizing and encouraging the government to provide actual socio-environmental accountability related to the socio-environmental problems. (KPA, 2017; LBH-Jakarta, 2015; LBH-Makassar, 2016; WALHI, 2015, 2016; Watchdoc, 2018). In fact, it is likely that these movements do not have enough political power for influencing the government policy directly at the national and local levels. It might be because these movements tend to be reactionary movements based on the one specific case (Darsono, 2015; KPA, 2017; LBH-Jakarta, 2015; LBH-Makassar, 2016; WALHI, 2015, 2016; Watchdoc, 2018). Consequently, the campaigns for advocating the socio-environmental accountability of the government are not nationally successful and become the main public concern (WALHI, 2016). Thus, it is expected that research on this issue can connect the people's and government's interest in the socio-environmental accountability concern (Fukuyama, 2014).

Furthermore, the other motivating point for doing the research within the government context is based on the critical view of Gray (2002), Tinker and Gray (2003), Saravanamuthu (2009), and Saravanamuthu and Lehman (2013). Their view promotes the development of an emancipatory socio-environmental accounting framework as the anti-thesis of market-based accounting that nowadays dominates the accounting discourse. It stands on the view that accounting practices should be shaped by the meaningful context-relevant accounts. It means the socio-locality context should have the main influence over the accounting structure and practice (Gray 2002; Tinker and Gray 2003; Saravanamuthu 2009; Saravanamuthu and Lehman 2013). This argument is the foundation of this research in studying the socio-environmental accountability alternative based on the socio-environmental activists' views in Indonesia and in trying to integrate the social-environmental-economic measurement to achieve a sustainable development within the society (Stiglitz et al., 2009; Sukoharsono, 2007).

1.3 Problem Statements

According to the discussion of the main issue above, it is believed that the local governments are likely to ignore the socio-environmental activists' views that promote the presence of the socio-environmental-economic (integrated) accountability within the local governmental bodies (KPA, 2017; LBH-Jakarta, 2015; LBH-Makassar, 2016; WALHI, 2015, 2016). Furthermore, in line with the encouragement of the accounting scholars, future research should focus on developing an emancipatory socio-environmental accounting framework that suitable with the local society's interest and environmental sustainability (Gray, 2002; Saravanamuthu, 2009; Saravanamuthu & Lehman, 2013; Stiglitz et al., 2009; Tinker & Gray, 2003). Therefore, this research aims to develop a new integrated accountability framework in the local governments based on the emancipatory socio-environmental accounting concept.

1.4 Research Objectives

This research explores the issue in a natural setting. Thus, the objective of this research is:

> This study is intended to develop a new integrated accountability framework in the local governments based on the emancipatory socio-environmental accounting idea.

1.5 Research Contributions

1. This study is expected to provide a significant theoretical and philosophical contribution to the recent development of accounting and accountability discourse. Concerning theories, this study fills the gap in the area of the emancipatory socio-environmental accounting framework discourse. Specifically, it contributes to the making of a new approach to develop an integrated accountability concept that focuses on managing the wealth of natural resources and the vulnerability of social-environmental problems based on socio-environmental reality in Indonesia. Philosophically, this study exposes more space to a discussion of another kind of critical view (philosophy). It is Tan Malaka's critical views that based on the socio-locality context of Indonesia.
2. In policy domain, specifically in the government bodies at the national and regional levels, this research could become a new base for developing the government integrated-accountability regulations and policy. Furthermore, it might also give a good indicator related to the relationship between the activists, society and government since the government accommodates the voices of civil society in socio-environmental accountability policy-making.
3. Practically, the results of this research could be new guidance to develop the government's integrated accountability system, specifically, in Indonesian local government bodies that face socio-environmental issues with their stakeholders. By using the results of this study, the study can help the governments to communicate their actions to the people easily and to reach a consensus related to the accountability for socio-environmental issues in Indonesia. Moreover, the study also provides a new perspective to the society about the government's accountability topics, and it can also be a global campaign tool for sustainable

development that is aligned with the public's interests, based on their respective socio-environmental issues.

1.6 Research Scope

In line with the objective of the research, this study is expected to develop a new integrated accountability framework for local governments grounded on the emancipatory socio-environmental accounting concept. It is based on the reality that local governments' accountability practices in Indonesia have a framework that refers to the president regulation Number 29/2014. This framework regulates the government's accountability practices from developing strategic plan to reviewing and evaluating performance. However, in reality, the implementation of the government accountability practices, specifically socio-environmental accountability, is widely questioned by the socio-environmental activists. Thus, this research will focus on discussing the current accountability framework with the socio-environmental activists' views for developing a new-suitable integrated accountability framework for local governments in Indonesia.

The framework developed in this research refers to a basic idea or conception structure (Merriam-Webster, 2018) of accountability activities in local governments as described in president regulation Number 29/2014.[7] Moreover, using the socio-environmental activists' ideas related to developing a new integrated accountability framework is grounded on the argument of Saravanamuthu and Lehman (2013) as they said, there is always a possibility to make a new account based on the society's views that can be used as a new accounting or accountability tools for solving a specific case.

References

BI. (2016). *Laporan Perekonomian Indonesia 2016*. Jakarta: Bank Indonesia.

Ball, A. (2004). A sustainability accounting project for the UK local government sector? Testing the social theory mapping process and locating a frame of reference. *Critical Perspectives on Accounting, 15*(4), 1009–1035.

Ball, A. (2005). Environmental accounting and change in UK local government. *Accounting, Auditing & Accountability Journal, 18*(3), 346–373. https://doi.org/10.1108/09513570510600738.

Ball, A. (2007). Environmental accounting as workplace activism. *Critical Perspectives on Accounting, 18*(7), 759–778.

Burrell, G., & Morgan, G. (1979). *Sociological paradigms and organisational analysis: Elements of the sociology of corporate life*. London: Heinemann.

Burritt, R. (2004). Environmental management accounting: Roadblocks on the way to the green and pleasant land. *Business Strategy and the Environment, 13*(1), 13–32.

[7] See Sect. 2.4 in this research for comprehensive explanation about the president regulation Number 29/2014 and the current accountability framework of governments in Indonesia.

Creswell, J. W. (2013). *Research design: Qualitative, quantitative, and mixed methods approaches*. Sage.
Darsono, F. (2015). *Koalisi Ornop Pasca Orde Baru: Studi tentang Jaringan Walhi dalam Kampanye Isu Hutan* (pp. 27–48). Masyarakat: Jurnal Sosiologi.
Deegan, C. (2002). Introduction: The legitimising effect of social and environmental disclosures—A theoretical foundation. *Accounting, Auditing & Accountability Journal, 15*(3), 282–311.
Diefenbach, T. (2009). New public management in public sector organizations: The dark sides of managerialistic "Enlightenment." *Public Administration, 87*(4), 892–909. https://doi.org/10.1111/j.1467-9299.2009.01766.x.
Djamhuri, A. (2009). *A case study of governmental accounting and budgeting reform at local authority in indonesia: An institutionalist perspective*. USM.
Fukuyama, F. (2014). *Political order and political decay: From the industrial revolution to the globalization of democracy*. New York: Farrar, Straus and Giroux.
Gray, R. (2001). Thirty years of social accounting, reporting and auditing: What (if anything) have we learnt? *Business Ethics: A European Review, 10*(1), 9–15. https://doi.org/10.1111/1467-8608.00207.
Gray, R. (2002). The social accounting project and accounting organizations and society privileging engagement, imaginings, new accountings and pragmatism over critique? *Accounting, Organizations and Society, 27*(7), 687–708. Retrieved from https://www.sciencedirect.com/science/article/pii/S0361368200000039.
Greenpeace. (2017). *About greenpeace*. Retrieved November 10, 2017, from https://www.greenpeace.org/international/en/about/.
Gunderson, L. H., Holling, C. S. (Eds.). (1995). *Barriers and bridges to the renewal of ecosystems and institutions*. New York: Colombia Press.
Hadiz, V. R. (2004). Decentralization and democracy in Indonesia: A critique of neo-institutionalist perspectives. *Development and Change, 35*(4), 697–718.
Harun, H., An, Y., & Kahar, A. (2013). Implementation and challenges of introducing NPM and accrual accounting in Indonesian local government. *Public Money and Management, 33*(5), 383–388. https://doi.org/10.1080/09540962.2013.817131.
KPA. (2017). *KPA Launching Catatan Akhir Tahun 2016*. Retrieved April 17, 2017, from https://www.kpa.or.id/news/blog/kpa-launching-catatan-akhir-tahun-2016/.
Kamayanti, A. (2016). *Metodologi Penelitian Kualitatif Akuntansi: Pengantar Religiositas Keilmuan*. Malang: Yayasan Rumah Peneleh.
Kamayanti, A. (2017). *Metodologi Konstruktif Riset Akuntansi: Membumikan Religiositas*. Malang: Yayasan Rumah Peneleh.
KataData. (2017). Industri Tambang Penopang Perekonomian. Retrieved February 8, 2018, from webpage website: https://katadata.co.id/infografik/2017/04/24/industri-tambang-penopang-perekonomian.
Kemenpan-RB. (2018a). *LaKIP Bukan Semata Laporan Pertanggungjawaban*. Retrieved May 21, 2018, from Kementrian Pendayagunaan Aparatur Negara dan Reformasi Birokrasi-RI website: https://www.menpan.go.id/site/berita-terkini/lakip-bukan-semata-laporan-pertanggung-jawaban.
Kemenpan-RB. (2018b). *Rapor SAKIP Membaik, 14 Pemda Mendapat DID*. Retrieved May 21, 2018, from Kementrian Pendayagunaan Aparatur Negara dan Reformasi Birokrasi-RI website: https://www.menpan.go.id/site/berita-terkini/rapor-sakip-membaik-14-pemda-mendapat-did.
LBH-Jakarta. (2015). *Atas Nama Pembangunan: Laporan Penggusuran Paksa di Jakarta Tahun 2015* (Vol. 2017). Retrieved from https://www.bantuanhukum.or.id/web/atas-nama-pembangunan-laporan-penggusuran-paksa-di-jakarta-tahun-201.
LBH-Makassar. (2016). *Siaran Pers Advokasi Masyarakat*. Retrieved from https://lbhmakassar.org/siaran-pers/.
Lapsley, I. (2009). New public management: The cruellest invention of the human spirit? *Abacus, 45*(1), 1–21. https://doi.org/10.1111/j.1467-6281.2009.00275.x.

MacLullich, K. K. (2004). Accounting, love and justice. *Accounting Auditing and Accountability Journal, 17*(3), 327–360.
Merriam-Webster. (2018). *Definition of framework*. Retrieved May 21, 2018, from merrian-webster online dictionary website: https://www.merriam-webster.com/dictionary/framework.
Niemann, L., & Hoppe, T. (2018). Sustainability reporting by local governments: A magic tool? Lessons on use and usefulness from European pioneers. *Public Management Review, 20*(1), 201–223. https://doi.org/10.1080/14719037.2017.1293149.
Osborne, R. (1972). *Who is the chairman of this meeting?: A collection of essays*. Toronto: Neewin.
Parker, L. (2005). Social and environmental accountability research: A view from the commentary box. *Accounting, Auditing & Accountability Journal, 18*(6), 842–860.
Qian, W., Burritt, R., & Monroe, G. (2011). Environmental management accounting in local government. *Accounting, Auditing & Accountability Journal, 24*(1), 93–128. https://doi.org/10.1108/09513571111098072.
Rajiani, I. (2011). A cultural ecology of new public management in Indonesia. *Journal of Administrative Science, 8*(1), 17–31.
Rasyid, M. R. (2003). Regional autonomy and local politics in Indonesia. *Local Power and Politics in Indonesia*, 63–72.
Republika Online. (2015). *Demo senin tuntut lestarikan hutan, cabut izin perusahaan pembakar lahan*. Retrieved from https://politik.rmol.id/read/2015/09/27/218798/demo-senin-tuntut-lestarikan-hutan-cabut-izin-perusahaan-pembakar-lahan.
Sanusi, A., Butler, B., & Rola-rubzen, M. F. (2013). Local politics and business in Indonesia: A non-market strategy approach. In *Provincial Governance Strengthening Programme Supported by UNDP Indonesia* (pp. 1–20). Unika Atmajaya, Jakarta: Unika Atmajaya.
Saravanamuthu, K. (2004). What is measured counts: Harmonized corporate reporting and sustainable economic development. *Critical Perspectives on Accounting, 15*(3), 295–302. https://doi.org/10.1016/S1045-2354(03)00063-7.
Saravanamuthu, K. (2009). A journey of socialising the risks associated with global warming: A Gandhian insight into Schumacher's total accounting and accountability. In *Extending Schumacher's Concept of Total Accounting and Accountability into the 21st Century* (pp. 121–189). https://doi.org/10.1108/S1041-7060(2009)0000014008.
Saravanamuthu, K., & Lehman, C. (2013). Enhancing stakeholder interaction through environmental risk accounts. *Critical Perspectives on Accounting, 24*(6), 410–437. https://doi.org/10.1016/j.cpa.2013.02.002.
Saravanamuthu, K., Brooke, C., & Gaffikin, M. (2013). ERP and precautionary ethics: Harnessing critical thinking to engender sustainability. *Journal of Information, Communication and Ethics in Society, 11*(2), 92–111. https://doi.org/10.1108/JICES-12-2012-0020.
Sari, L. P., Djamhuri, A., & Kamayanti, A. (2014). Accrualization in the public sector: The case of Situbondo district. *The International Journal of Accounting and Business Society, 22*(2), 55–68.
Shahib, H. M., & Risky, F. R. (2017). Accountability in the Internet Era: A lesson from local governments in Indonesia. *Hasanuddin Economics and Business Review, 1*(1), 57–74.
Shahib, H. M., & Irwandi, S. A. (2016). Violation regulation of financial services authority (FSA), financial performance, and corporate social responsibility disclosure. *Journal of Economics, Business, and Accountancy Ventura, 19*(1), 141–154. https://doi.org/10.14414/jebav.v19i1.533.
Stiglitz, J. E., Sen, A., & Fitossi, J.-P. (2009). *Commission on the measurement of economic performance and social progress report-2009*. Paris: The Commission on The Measurement of Economic Performance and Social Progress-France.
Sukoharsono, E. G. (2005). Environmental accounting: A map of environmental regulations and policies of the Indonesian Central Government. *The International Journal of Accounting and Business Society, 13*(2), 1–22.
Sukoharsono, E. G. (2007). Green accounting in Indonesia: Accountability and environmental issues. *The International Journal of Accounting and Business Society, 15*(1), 21–60.

Timur, T. (2015). *Pemerintahan Jokowi harus lindungi ekosistem karst yang terancam industri pertambangan*. Retrieved from https://www.tribunnews.com/regional/2016/01/11/pemerintahan-jokowi-harus-lindungi-ekosistem-karst-yang-terancam-industri-pertambangan.

Tinker, A., & Gray, R. (2003). Beyond a critique of pure reason: From policy to politics and praxis in environmental and social research. *Accounting, Auditing, and Accountability Journal, 16*(5), 727–761.

WALHI. (2015). *Tinjauan Lingkungan Hidup 2015 WALHI*. Retrieved from https://walhi.or.id/wp-content/uploads/2016/12/OutLook2015_W.pdf.

WALHI. (2016). *Tinjauan Lingkungan Hidup WALHI 2016: Keharusan Pembenahan Struktural untuk Perbaikan Tata Kelola*. Retrieved from Wahana Lingkungan Hidup Indonesia website: https://www.walhi.or.id/2016/12/22/outlook-2016-tujuan-lingkungan-hidup/.

Watchdoc. (2018). *Kumpulan film dokumenter kerusakan lingkungan di Indonesia*. Retrieved February 14, 2018, from Youtube website: https://www.youtube.com/user/watchd0c/featured.

Williams, B., Wilmshurst, T., & Clift, R. (2011). Sustainability reporting by local government in Australia: Current and future prospects. *Accounting Forum, 35,* 176–186.

World-Bank. (2012). *Doing business in Indonesia 2012*. Washington, DC: World-Bank.

Chapter 2
Literature Review

There is nothing eternal but the change.
—Heraclitus (an ancient Greek philosopher)

2.1 Accountability Over Time

The term of accountability is embedded in accounting discourses. It can be said that the accounting practice is a consequence of accountability idea otherwise both have firmly correlated each other. This can be seen from the history of accounting and accountability words that have a historical basis in record keeping actions related to money-transactions and governance activities of the business, kingdom, empire, and state (Ezzamel, 1997; Mulgan, 2000; Seidman, 2004).

Even now, the historical and philosophical basis of the accountability term still can be traced in today's discourse and organizational practices. Especially, if it is linked to the definitions of accountability from the policymaker and think-tank institutions. For example, Organization for Economic Co-operation and Development (OECD)[1] stated:

> Accountability is a key concept in modern management theory and practice. It means that managers [entities] are held responsible for carrying out a defined set of duties or tasks, and for conforming with rules and standards applicable to their posts (OECD, 2018).

Moreover, AccountAbility[2] and KNKG[3] explain that:

---

[1]Organization for Economic Co-operation and Development (OECD) is an inter-developed-governmental economic organization that focuses on democracy and market economy cooperation founded in 1961.

[2]AccountAbility is a world nonprofit institution with representatives in London, Washington, Beijing, Geneva, Sao Paulo and San Francisco, established in 1995 to promote accountability innovations that advance sustainable development.

[3]*Komite Nasional Kebijakan Governance (KNKG)* is an Indonesia's committee that focuses on assisting the private and public agencies in facing and solving the governance issues.

H. M. Shahib, *Towards the Local Government's Integrated Accountability Framework*,
https://doi.org/10.1007/978-981-33-6617-6_2

> AccountAbility is acknowledging, assuming responsibility for and being transparent about the impacts of your policies, decisions, actions, products and associated performance. It obliges an organization to involve stakeholders in identifying, understanding and responding to sustainability issues and concerns, and to report, explain and be answerable to stakeholders for decisions, actions, and performance (AccountAbility, 2008). Or an explanatory function and responsibility action of an organization's activities [in private and public area] (KNKG, 2008).

Merriam-webster dictionary (Merriam-Webster, 2018) simplifies these meanings of accountability as follows:

> Accountability (noun) is the quality or state of being accountable; especially: an obligation or willingness to accept responsibility or to account for one's actions.

According to that simple definition, it can be seen that the accountability concept and practice nowadays is built on three points of those definitions. The first point is "*the quality or state of being accountable*". The second point is "*the obligation or willingness to accept responsibility*". The third point is "*to account for one's actions*". These three points are further discussed in the next sections.

1. *Accountability-The quality or state of being accountable*

Accountability in terms of "the quality or state of being accountable" is related to the moral and ethics of a person and an organization in their social interaction (Dykstra, 1939; Mulgan, 2000; Williams, 2006). At this point, the meaning of accountability is similar to the meaning of social responsibility, which is the individual and organization entity interacting socially within the universal ethics framework. For instance, the person and organization should act honestly, answerability, trustworthy, responsibly, and blameworthy in their social interactions (Dykstra, 1939; Mulgan, 2000; Williams, 2006). Moreover, related to the organization's ethical actions, all activities in organizations, public or private, should reflect the universal ethical value in the organization's daily activities, starting from the organization's policy, fabrication of goods or services, governance, and administration (Dwivedi, 1989; Williams, 2006).

A question then arises; how can the public assure the quality of accountability of a person and an organization? There are two main answers for this question. Firstly, the public can assure the ethical actions by directly seeing and feeling a person and an entity's acts in daily activities (Dwivedi, 1989; Schedler, Diamond, & Plattner, 1999; Williams, 2006). Secondly, the public can be convinced by using a report about the person's and entity's main activities (Ezzamel, 1997; Gray, Owen, & Adams, 1996; Macintosh, 2002). For example, there is a report of police record (in Indonesia: *SKCK*[4]) indicating that an individual is free (not free) from criminal actions. Furthermore, there is a tax report that clarifies that an individual is free (not free) from tax debts. For entity, a report associated with the entity's operations is widely known as accountability report (AccountAbility, 2008; Ezzamel, 1997; Gray et al., 1996; Macintosh, 2002; Ramanna, 2013), for instance, operational report, financial report, audit report, and marketing report.

---

[4] *Surat Keterangan Catatan Kepolisian (SKCK).*

2. *Accountability-The obligation or willingness to accept responsibility*

With regard to the second point of accountability definition, that is "the obligation or willingness to accept responsibility", it is interesting to revisit the opinion of Dwivedi (1989) and Macintosh (2002) that the accountability can be shaped by the social structure of the society. The "obligation or willingness to accept responsibility" is a proof that there is a social structure and a social interaction behind the accountability (Dwivedi, 1989; Macintosh, 2002). Dwivedi (1989) proposes that there are eight kinds of accountability i.e. moral, political, administrative, managerial, electorate relation, market, legal or judicial, and professional accountability. However, in the current context, it can simply be classified into four kinds of accountability namely: ethics accountability (moral, professional, and legal or judicial), governance accountability (administrative, managerial, and market), and political accountability (electorate and political relation).

Various types of accountability exist as follows: (1) The ethics accountability can be felt in the daily social interaction and can be found in the ethics report in public and private entities (Dwivedi, 1989; Weber & Wasieleski, 2013). (2) The governance accountability in firms and local governments can be sensed in their service and goods, and can be seen in their annual reports (Gray et al., 1996; KNKG, 2008; Stanton & Stanton, 2002). (3) The political accountability can be perceived in the government's policy implementation and report (Adsera, Boix, & Payne, 2003; Boyne & Law, 1991; Mulgan, 2000).

3. *Accountability-To account for one's actions*

According to the previous explanations, generally, accountability is likely to be perceived as a responsible action as well as a kind of a responsibility report. In other words, it can be said that the accountability report should stand on the reality of accountability actions (Boyne & Law, 1991; Ezzamel, 1997; Gray et al., 1996; Macintosh, 2002; Mulgan, 2000).

In line with these arguments, it is interesting to see the last part of the definition of accountability in terms of "to account for one's actions". This definition can explain what happens in the world today. In line with the fast-growing economy and technology nowadays, the interaction between person and person, person and the society, person and business firm, person and government agencies, and vice versa become faster and more complex (Fukuyama, 2006, 2014). Thus, for making the interactions simpler, the accountability actions today tend to be perceived as a kind of responsibility report (AccountAbility, 2008; Adams & McNicholas, 2007; Gray et al., 1996; Mulgan, 2000). Individual or organization is called an accountable entity if they have a responsibility report in their activities (Boyne & Law, 1991; Gray, Bebbington, & Collison, 2006; Mulgan, 2000).

As a result, currently, many forms of accountability report are present. For instance, financial, compliance, operational, annual and sustainability reports are available in private and public organizations (AccountAbility, 2008; Boyne & Law, 1991; Davis, Guenther, Krull, & Williams, 2013; Gray et al., 1996, 2006; Hockings,

Cook, Carter, & James, 2009). The phrase "to account for one's actions" tends to be a basis for explaining the situations today widely open for the development of many new accountability report models as long as they have a root in the social reality (Macintosh, 2002).

2.2 Accountability Reports as the Form of Socio-Politic Legitimacy

In line with the previous explanation that the accountability actions in private and public entities nowadays are symbolized by various accountability reports (Boyne & Law, 1991; Davis et al., 2013; Gray et al., 1996, 2006; Hockings et al., 2009). The socio-politic legitimacy discourse is then useful for elaborating this issue more deeply. The accountability action and types of its reports are not present from a vacuum. Rather, they are influenced by sociopolitical conditions, known as the socio-political legitimacy process (Deegan, 2002; Gray et al., 1996; Guthrie & Parker, 1989; Islam & Deegan, 2010; Macintosh, 2002).

In general, the socio-political legitimacy discourse proposes that there is always a respected value in a society, which is present from the culture, sociopolitical interactions, and the agreement between all entities in the society (Archel, Husillos, Larrinaga, & Spence, 2009; Fukuyama, 2014; Patten, 1991). Consequently, all bodies should respect and obey this value if they still want to be considered and legitimized as a part of the social community. Otherwise, the society will give them punishment and/or revoke their legitimacy (Deegan, 2007; Fukuyama, 2014; Lanis & Richardson, 2013).

Initially, the act of respecting the community's rules and values was believed sufficient for maintaining the legitimacy of a social body (Dowling & Pfeffer, 1975; Dykstra, 1939; Mulgan, 2000; Williams, 2006). However, as time goes on and in line with the fast-revolution in the economy and technology that change social interactions between people and organizations, making them more complex, doing well is believed to be inadequate (Gray et al., 1996; Gray, 2002). It is understood that the entity should communicate the summary of their good and bad activities to the public in the form of reports (Boyne & Law, 1991; Davis et al., 2013; Gray et al., 1996, 2006; Hockings et al., 2009). Therefore, the society can discuss and assure all activities of an organization, starting from upstream to downstream or vice versa. At the end, the society will support the entity's presence (Archel et al., 2009; Koppell, 2008; Lanis & Richardson, 2013).

For example, nowadays, based on the financial regulation, the public can check and re-check the annual financial and operation information of a business entity by reading its financial and annual report. In this case, the financial and annual reports become the accountability tools of a profit organization for maintaining its legitimacy (Lanis & Richardson, 2012, 2013).

Moreover, as can be seen from the above example, the social characteristics greatly affect the accountability report form. The "regulation" is a keyword to understand this situation. Today, all accountability reports, both in private and public sectors, are bound to the regulation (Cameron, 2004; Vogel, 2010), and the regulation is the result of political agreements between societies, government, and regulatory bodies (Benmelech & Moskowitz, 2010; Fukuyama, 2014; Radaelli, 2005). This situation then becomes a philosophical basis of the presence of accountability reports in many kinds of the profit and public agencies.

Therefore, it is interesting to see what happens these days, from the theoretical discourse to practical domain of accountability. As can be seen today, in the private field, the accountability mechanisms turn into more advances. Private companies should prepare many accountability instruments such as financial reports, annual reports, corporate social responsibility (CSR) reports, and sustainability reports since the discourse of sustainable development (Elliott, 2012; Mebratu, 1998) and triple bottom line (profit, people, planet) (Elkington, 1998) turn out to be the main discourse and are believed by the societies and regulatory bodies around the world (Jennifer Ho & Taylor, 2007; Tschopp, 2005). As the effect of that discourse domination, today, most of private companies around the world produce CSR reports and account their business operations in many forms of accountability reports for ensuring that their business actions are in line with the public attention on the sustainable development paradigm (Adeyeye, 2011; Islam & Deegan, 2010; Jindrichovska & Purcarea, 2011; Tallontire, 2007).

However, in the government field, especially local governments in middle-income countries like Indonesia, the market paradigm based on monetary measurement tends to dominate the accountability practices (Diefenbach, 2009; Djamhuri, 2009; Lapsley, 2009; Rajiani, 2011) although the present rules and accountability instruments support the sustainable development.[5] This situation can be seen from the fact that many Indonesian socio-environmental NGOs actively criticize the government's socio-environmental accountability, especially the local governments (KPA, 2017; LBH-Jakarta, 2015; LBH-Makassar, 2016; WALHI, 2015, 2016; Watchdoc, 2018).

The NGOs believe that local governments tend to focus only on the economic growth achievement (KPA, 2017; LBH-Jakarta, 2015; LBH-Makassar, 2016; WALHI, 2015, 2016; Watchdoc, 2018). They lose the environmental endurance, creating social problems in the society. For instance, as a consequence of the easy investment in natural resources sectors (metrotvnews.com, 2017; Sindonews.com, 2016), the number of natural resource conflicts between indigenous people and firms as well as local people being banished from their home and cultivated fields increase (KPA, 2017; LBH-Jakarta, 2015; LBH-Makassar, 2016). Moreover, as a consequence of the lack of environmental accountability, the problems of forest fires, local water-food scarcity and ecosystems destruction (WALHI, 2015, 2016; Watchdoc,

[5](1) The Law of Environment Number 32/2009. (2) The President Regulation Number 29/2014 regarding The Accountability System of Government agencies' performance. (3) The Regulation of the Minister of State-Apparatus Employment *(Permenpan)* Number 53/2014 regarding The Guidelines of Performance Accountability report in the Government Agencies.

2018) emerge and are difficult to be controlled (KPA, 2017; LBH-Jakarta, 2015; LBH-Makassar, 2016; RMOL.ID, 2015; Timur, 2015; WALHI, 2015, 2016).

Therefore, it is interesting to discuss more deeply, what is really happening in the local government's accountability in Indonesia? Especially, when it is related to the socio-political construction of the accountability instrument in this sustainable development paradigm era that has become a worldwide concern (Adeyeye, 2011; Islam & Deegan, 2010; Jennifer Ho & Taylor, 2007; Jindrichovska & Purcarea, 2011; Tallontire, 2007; Tschopp, 2005).

2.3 Accountability, NPM, and Sustainable Development in Local Government in Indonesia

In the end of 1990s, the government's accountability and transparency discourse started to become the main public focus in Indonesia (Hill & Shiraishi, 2007; Kristiansen, Dwiyanto, Pramusinto, & Putranto, 2009) due to the 1997–1998 economic crisis (Cole & Slade, 1998) and the corruption, collusion, and nepotism (*KKN*) conducted massively by President Soeharto and his cronies for 32 years (McLeod, 2000). After the 1998 reform movement succeeded in overthrowing President Soeharto (Aritonang, 1999), the Indonesia Government began to fix the state administration and tried to present accountable and transparent governance (Hill & Shiraishi, 2007; Kristiansen et al., 2009). In line with that policy, the government also implemented political and economic decentralization policy, known as *otonomi daerah* (regional autonomy) (Hadiz, 2004; Rasyid, 2003; Sanusi, Butler, & Rola-rubzen, 2013) in order to give more space to local governments to manage their area and people by themselves (Rasyid, 2003). As a consequence of this regional autonomy policy, the local governments should be accountable and transparent for maintaining their legitimacy in the society (Eckardt, 2008).

Some regulations are then designed in the reform era with the purpose of supporting the government policies to be more accountable and transparent, especially in financial sector. Some of these rules include:

(1) Law Number 17/2003 regarding State Finance.
(2) Law Number 1/2004 regarding State Treasury.
(3) Law number 15/2004 regarding the Audit of State Financial Management and Accountability.
(4) Government Regulation Number 24/2005 regarding the Government Accounting Standard (cash toward accrual basis).
(5) Government Regulation Number 71/2010 regarding The Government Accounting Standard (accrual basis).

Based on these regulations, the central and local governments are obliged to make financial statements that consist of the statement of Budget realization (LRA), changes in Budget Balance (SAL), changes in equity (LPE), Balance Sheet *(Neraca)*,

Operational Statement (LO), Cash Flow Statement (LAK), and notes of financial statement (CaLK).[6] Moreover, the consequence of the financial statements is that government institutions should be annually audited by the Audit Board of the Republic of Indonesia (BPK-RI) at the end of fiscal year (Julianto, 2010).[7]

Besides regulations in the financial sector, after the reform, there are also many new regulations for ensuring that the central and local governments in Indonesia can achieve accountability and transparency. For instance:

(1) The regulation of the Head of the State Administration Institution (LAN) Number 239/IX/6/8/2003 regarding the Improvement of the Guidelines of Accountability Reporting Preparation in the Government Agencies.
(2) President Regulation Number 29/2014 regarding the Accountability System of Government agencies' performance.
(3) Regulation of the Minister of State-Apparatus Employment (*Permenpan*) Number 29/2010 regarding the Guidelines of Performance Determination Compilation; and Performance Accountability report in Government Agencies (the latest revision of *Permenpan* Number 53/2014).

These regulations then generate a report called the Performance Accountability Report of Government Agencies (LAKIP).[8] The public should appreciate government's regulations that support the accountability and transparency in the government sector. However, referring to the accountability terms used by the government, such as the accrual-based financial statements, cash-flow statement, balance sheet, operational statement, equity statement, accountability report, and audit opinion, there is an indication that the New Public Management (NPM) paradigm significantly influences the regulation process which binds the government's accountability activities in Indonesia (Djamhuri, 2009; Rajiani, 2011). This view slowly affects the government's mindset in the policy-making, control, and accountability practice to become extremely monetary-based and market-oriented (Diefenbach, 2009; Djamhuri, 2009; Rajiani, 2011). Because of these reasons, many scholars criticize the NPM approach. They believe that government organizations have a main duty to serve the society. Thus, the monetary-based and market-orientation views are not appropriate as the government organizations' philosophy (Apple, 2005; Diefenbach, 2009; Djamhuri, 2009; Lapsley, 2009; Maesschalck, 2004).

Historical facts show that in the Indonesian context, NPM practices started when the economic crisis hit Indonesia between 1997 and 1998. As a consequence, the Indonesian Government tried to reform its birocration and accepted the letter of intent

---

[6]In Indonesian Language: *Laporan Realisasi Anggaran (LRA), Laporan Perubahan Saldo Anggaran Lebih (SAL), Laporan Perubahan Ekuitas (LPE), Neraca, Laporan Operasional (LO), Laporan Arus Kas (LAK), dan Catatan atas Laporan Keuangan (CaLK).*

[7]There are five audit opinions issued by BPK-RI: (1) *Wajar tanpa pengecualian (WTP)*/unqualified opinion, (2) *Wajar tanpa pengecualian dengan paragraf penjelasan*/unqualified opinion with explanatory paragraph, (3) *Wajar dengan pengecualian (WDP)*/qualified opinion, (4) *tidak wajar*/adverse opinion, (5) *menolak memberikan opini*/disclaimer opinion (Julianto, 2010).

[8]*Laporan Akuntabilitas Kinerja Instansi Pemerintah (LAKIP).*

of the debt-aid from IMF (International Monetary Fund), which required the government practices to adopt the IMF policies (Hadi et al., 2012; Mulawarman, 2016), that tend to market-oriented or neoliberism-oriented (Ramli & Nuryadin, 2007). Consequently, as explained in the previous paragraphs, there are many regulations that support the market-oriented of IMF policies, specifically reducing the government's expenditure (Robison & Rosser, 1998). This situation made the government institutions [consciously or unconsciously], slowly changed to use NPM styles as a consequence of neoliberalism agenda behind the IMF policies (Harun, An, & Kahar, 2013).

According to the Diefenbach (2009), the government institutions that have implemented the NPM approach will have certain characteristics similar to those of private organizations. It is based on the philosophical basis of the NPM itself, which tries to implement the private-firms' management and organization style to government agencies. Therefore, many private organizations' terms such as performance, costs, effective-efficiency, and audited-oriented work can be found in the government organizations. These conditions then slowly affect the government actions. The governments that applied the NPM tend to act like profit-oriented business firms. They work as effective and efficient by minimizing time and cost. They are also market-oriented. For example, they focus on the investment and economic growth, and do not concentrate on non-monetary matters such as the environmental, social, and culture issues (Diefenbach, 2009; Djamhuri, 2009; Lapsley, 2009; Rajiani, 2011).

These situations occur in all regions in Indonesia, from the central government to the local governments. The NPM and its forms, especially the accountability and transparency instruments in the government institutions, such as financial statement (LK), and performance accountability report of government agencies (LAKIP) become the justification tools. It means that they just act as the symbol of the presence of good government governance (GGG) in the government institutions. Like business companies, the government agencies, especially local governments, tend to be proud of the rupiah (Rp) numbers listed on their financial statements and the audit opinion of their financial statements (Kompas, 2017).

Moreover, on the international business side, it is believed that the existence of the modern government accountability instruments, which is based on the NPM philosophy, can invite many investors and business transactions to come to Indonesia (Grindle, 2004; Henisz & Zelner, 2005; World-Bank, 2012). Responding to these situations, the central and local governments welcome them by presenting the soft investment regulations easily complied by the local and foreign investors. The central and local governments expect that this policy can help them achieve the income target and maintain the national and regional economic growth (metrotvnews.com, 2017; Sindonews.com, 2016).

Despite the positive effects on the Gross Domestic Product (GDP), the investment opportunities based on the government's accountability monetary-information also have negative effects on the environmental sustainability and social stability of the small communities (Sanusi et al., 2013; World-Bank, 2012). In many regions, when the local governments open the door for investments, specifically investments in the natural resource sectors, socio-environmental problems emerge. Conflicts related to natural resources between indigenous people and firms, local people being banished from their home and cultivated fields (KPA, 2017; LBH-Jakarta, 2015; LBH-Makassar, 2016), forest fires, local water-food scarcity and ecosystems destruction are some of the problems (WALHI, 2015, 2016).

Unfortunately, these socio-environmental issues are not adequately covered by the government's accountability mechanisms. It is likely that the government's accountability instruments only follow the NPM philosophy that tends to focus just on monetary and market value standards (Diefenbach, 2009). As the evidence, besides the socio-environmental negative impacts which happened, there are massive protests socio-environmental NGOs against the government's socio-environmental accountability in many regions (KPA, 2017; LBH-Jakarta, 2015; LBH-Makassar, 2016; RMOL.ID, 2015; Tribun Timur, 2015; WALHI, 2015, 2016).

Observing this situation, there seems to be a fracture and gap of socio-political legitimacy in the accountability practices in Indonesia. Although, historically, the Government of Indonesia always takes an important part in the ratification of the international socio-environmental convention held by the United Nations, starting from the 1972 Stockholm conference to 17 sustainable development agendas (Sugandhy & Hakim, 2007; Waluyo, 2004). All of these international socio-environmental agreements inspire all socio-environmental laws and policies in Indonesia (Sugandhy & Hakim, 2007; Waluyo, 2004). For example, Indonesia has Law Number 40/2007 regarding Limited Companies, which governs the Limited Companies to practice in frame of sustainable development goals. Moreover, annually, the firms should make a corporate social responsibility (CSR) report or sustainability report as a sign of firm's accountability that supports sustainable development goals. In addition, Indonesian firms tend to use GRI standards (GRI, 2016; Shahib & Irwandi, 2016; Sukoharsono, 2007) as a benchmark for preparing their sustainability or CSR report. More explicitly, Indonesia also has Law Number 32/2009 regarding the Protection and Management of Environment that administrates environmental issues related to public and private institutions activities.

Thus, it can be said that Indonesia has implemented the sustainable development goals (UNDP, 2018)[9] in its governance practice in private and public sectors. However, many negative socio-environmental impacts happen, and there are massive protests that come from socio-environmental NGOs associated with the government's socio-environmental activities and accountability in many regions (KPA, 2017; LBH-Jakarta, 2015; LBH-Makassar, 2016; RMOL.ID, 2015; Timur, 2015; WALHI, 2015, 2016). Therefore, the implementation of sustainable development goals can be questioned and reviewed again specifically by using socio-politic legitimacy lens.

[9]Sustainable Development Goals (SDGs) is a worldwide demand that issued by United Nations to do some acts to end poverty, protect the earth, and guarantee that all humankind live in prosperity and peace. Currently, 17 goals linked each other: (1) No Poverty. (2) Zero Hunger. (3) Good Health & Well-being. (4) Quality education. (5) Gender equality. (6) Clean water and sanitation. (7) Affordable and clean energy. (8) Decent work and economic growth. (9) Industry, Innovation, & infrastructure. (10) Reduced inequalities. (11) Sustainable cities and communities. (12) Responsible consumption and production. (13) Climate action. (14) Life below water. (15) Life on land. (16) Peace, justice & strong institutions. (17) Partnerships for the goals (UNDP, 2018).

Based on these arguments, it is intriguing to understand how the government's accountability practices have been implemented in order to identify the socio-legitimacy gap leading to the massive protests of the NGOs against the local government's social-environmental accountability in Indonesia.

2.4 Local Government Accountability in Indonesia

The Indonesian Government's accountability in the form of reports has been started since the presence of Presidential Instruction Number 7/1999 regarding the Performance Accountability of Government Agencies, meeting the 98-reform movement which required the government to become more accountable and transparent (Hill & Shiraishi, 2007; Kristiansen et al., 2009).

Many changes occurred in Indonesia's accountability discourse, which are caused by economic and socio-political changes from inside and outside Indonesia. The Government then adjusted the rules of accountability instruments in the context of the present time by releasing President Regulation Number 29/2014 regarding the Accountability System of Government agencies' performance and the Regulation of the Minister of State-Apparatus Employment (*Permenpan*) Number 29/2010 regarding the Guidelines of Performance Determination. These regulations then generate a report called the Performance Accountability Report of Government Agencies (LAKIP/LKIP).[10]

Generally, both regulations complement each other. The President Regulation Number 29/2014 is a legal basis of the system of performance accountability in government agencies (*SAKIP*)[11] which includes a systematic sequence of activities, tools, and procedures designed for the purpose of setting and measuring, data collection, classification, resuming, and reporting performance at government agencies in order to be accountable and to improve performance in government agencies, from central to local regions in Indonesia. Furthermore, the Regulation of the Minister of State-Apparatus Employment (*Permenpan*) Number 29/2010 acts as a guideline to prepare the accountability reports as an integral part of the government accountability system.

In practice, the implementation of *SAKIP* must be in harmony with the administration of the Government Budget and Accounting System, control activities, and evaluation of the development plan implementation. According to the President Regulation Number 29/2014 Chap. 5, the application of *SAKIP* in government agencies covers: (1) the 5 years strategic plan[12]; (2) performance agreement; (3) performance

---

[10]*Laporan Akuntabilitas Kinerja Instansi Pemerintah (LAKIP)/Laporan Kinerja Instansi Pemerintah (LKIP).*

[11]*Sistem Akuntabilitas Kinerja Instansi Pemerintah (SAKIP).*

[12]There are two strategic plans: (1) RPJMD: *Rencana Pembangunan Jangka Menengah Daerah* (regional development planning document for a period of 5 (five) years. (2) Renstra-PD (the strategic plan of government's agencies based on RPJMD).

measurement; (4) performance data management; (5) performance reporting; and (6) review and performance evaluation. Furthermore, The performance indicators used in SAKIP should meet the following criteria: (a) specific; (b) measurable; (c) attainable; (d) time bound; and (e) traceable. Simply, the implementation of *SAKIP* creates the following framework (Fig. 2.1).

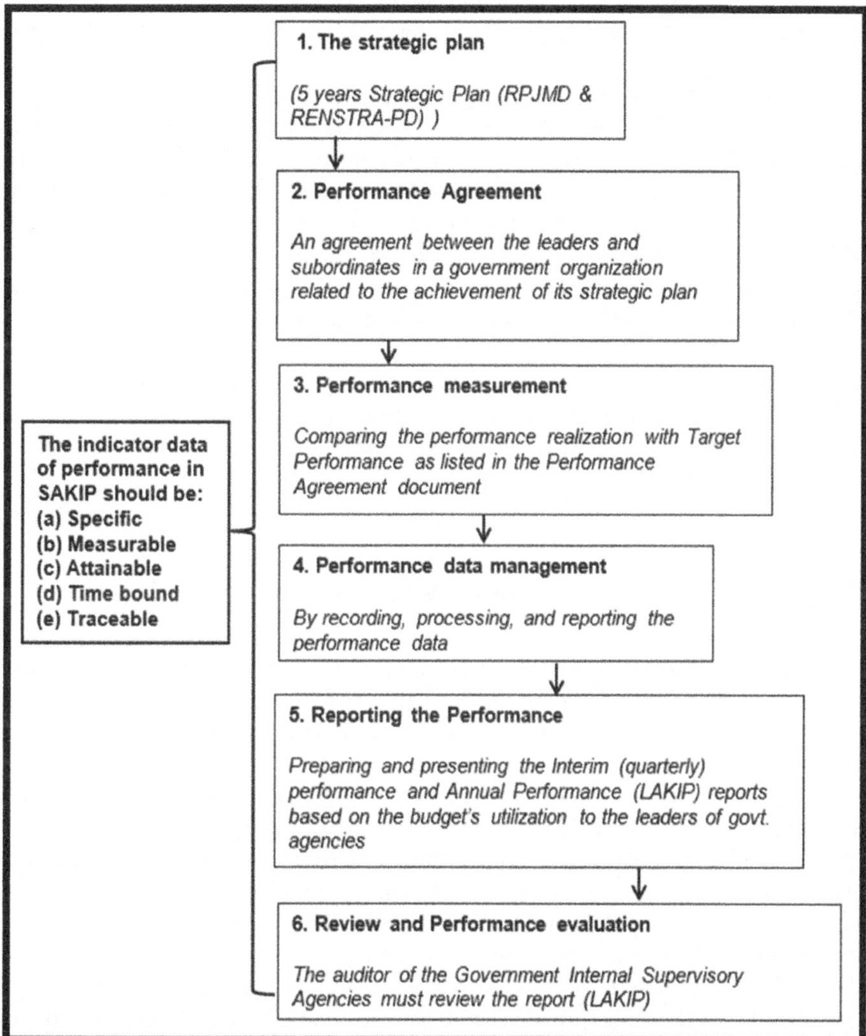

Fig. 2.1 The framework of accountability performances system in government agencies (*SAKIP*) (developed based on the President Regulation Number 29/2014 regarding the accountability system of government agencies' performance)

Along with the implementation of *SAKIP*, the Indonesian government agencies, from central to local regions, enter a new era of accountability. All of them should act according to the *SAKIP*'s framework and be accountable for what they have done. Therefore, a question then arises; what really happens with the implementation of this accountability framework. That there are massive protests from NGOs and societies against the local governments' socio-environmental accountability in many regions of Indonesia should be taken into account (KPA, 2017; LBH-Jakarta, 2015; LBH-Makassar, 2016; RMOL.ID, 2015; Tribun Timur, 2015; WALHI, 2015, 2016; Watchdoc, 2018).

According to the socio-political legitimacy perspective, the protests from the socio-environmental NGOs are a sign of a crack in the socio-political legitimacy status (Fukuyama, 2014) particularly related to the government's socio-environmental accountability issues in Indonesia. An interesting opinion related to the way to fix the socio-political legitimacy condition is proposed by Fukuyama (2014). He considers that if there is a problem (protest from the people) linked to the government's socio-political policy, the people should try to review the legitimacy basis of that policy for ensuring that it can be repaired or changed by the society itself. Moreover, Fukuyama (2014) believes that a scholar's direct study can be an effective way to identify the real problem for fixing the gap in the socio-political legitimacy circumstances because a research approach can give a clear insight into and recommendations for an issue. These statements then become another basis for doing this research.

2.5 Review of Socio-Environmental Accounting Studies

The statements of Fukuyama (2014) in the previous paragraph align with the concern of socio-environmental accounting studies in the last three decades. Many accounting scholars, especially those who concern about socio-environmental issues, believe that the socio-political legitimacy process plays an important role in discussing, creating and criticizing an accounting or accountability model (Deegan, 2002; Gray et al., 1996; Guthrie & Parker, 1989; Islam & Deegan, 2010; Macintosh, 2002).

This can be seen from the development and implementation of corporate responsibility report (CSR) and sustainability report in the present day. There is a long discussion, even political lobbying, between the scholars, societies (including NGOs), regulatory bodies, and the firms for taking an agreement of what kind of the regulation and report which must be implemented in order to overcome the socio-environmental accountability issues in the private sector (Adeyeye, 2011; Gray, Owen, & Adams, 1996; Gray, 2001; GRI, 2002; Stiglitz, Sen, & Fitossi, 2009; Sukoharsono, 2007; Tallontire, 2007).

In the same way, this situation can be a reference for assessing the position of social-political legitimacy within the government's accountability context, especially in Indonesia where many socio-environmental NGOs are questioning and protesting the social-environmental accountability practices in the local governments (KPA

2017; LBH-Jakarta, 2015; LBH-Makassar, 2016; RMOL.ID, 2015; Timur, 2015; WALHI, 2015, 2016; Watchdoc, 2018). The academic engagement in research is one of socio-political discussion tools (Fukuyama, 2014) which can be chosen as a method to evaluate and develop the legitimacy basis of accountability practices of the local governments in Indonesia.

Moreover, referring to studies of socio-environmental accounting and accountability issues that rely on the local government cases, there are still limited studies focusing on socio-environmental accountability forms in the local government agencies that directly involved in managing the social and environmental issues (Ball, 2004, 2005, 2007; Joseph, Pilcher, & Taplin, 2014; Niemann & Hoppe, 2018; Qian, Burritt, & Monroe, 2011; Shahib & Risky, 2017; Williams, Wilmshurst, & Clift, 2011). Specifically, in the Indonesian setting where the issues of corporate social responsibility report (CSR) of private firms listed on the Indonesia Stock Exchange (BEI) are likely to dominate the discourse of socio-environmental accounting and accountability studies at this present time (Shahib, 2016; Sukoharsono, 2007).

Nevertheless, there is a special motivation to conduct an in-depth study on the socio-environmental accountability themes. It originates from the study of Saravanamuthu and Lehman (2013) who encourage researchers to conduct further studies related to the basic theory and measurement improvement of the socio-environmental accountability that fit with the "present" and "locality" settings. Saravanamuthu and Lehman (2013) used the critical view of Gray (2002), Saravanamuthu (2009), Tinker and Gray (2003) in developing an emancipatory socio-environmental accounting framework as the anti-thesis of market-based accounting that nowadays dominates the accounting discourse.

The emancipatory socio-environmental accounting framework of Saravanamuthu and Lehman (2013) established the ideas that there is always a possibility to make a new account based on the society's views that can be used as a new accounting tool for solving a specific case. For example, the study of Saravanamuthu and Lehman (2013) developed an environmental accounting matrix for helping the irrigators manage the catchment. Therefore, it can also be said that the development of the emancipatory socio-environmental accounting framework depends on the socio-locality contexts that significantly influence the accounting structure and practice (Gray, 2002; Tinker and Gray, 2003; Saravanamuthu, 2009; Saravanamuthu and Lehman, 2013).

This argument becomes the foundation of this research for fixing the socio-political legitimacy situation related to the socio-environmental accountability issues. Especially, that is associated to the regional governments in Indonesia that these days their socio-environmental accountability practices are still greatly questioned by the local socio-environmental NGOs, society, and academicians (KPA, 2017; LBH-Jakarta, 2015; LBH-Makassar, 2016; Sukoharsono, 2007; WALHI, 2015, 2016; Watchdoc, 2018). It is expected that, by using the new emancipatory socio-environmental accounting framework, this research can integrate the social-environmental-economic measurement as a basis for developing a new integrated accountability framework for local governments in order to achieve sustainabile development in the Indonesian society (Stiglitz et al., 2009; Sukoharsono, 2007; UNDP, 2018).

2.6 Current Research Setting

In line with the previous explanations, it is believed that the socio-political-locality contexts have a strong/significant influence on the accounting structure and practice (Gray, 2001; Gunderson & Holling, 1995; Macintosh, 2002; MacLullich, 2004; Parker, 2005; Saravanamuthu, 2004, 2009; Saravanamuthu & Lehman, 2013; Stiglitz et al., 2009; Sukoharsono, 2007, 2005). Hence, ontologically, the basic idea of accounting and accountability tools are classified as the idealism or the constructed reality (Burrell & Morgan, 1979; Creswell, 2013; Macintosh, 2002).

Based on this reason, this research will use a natural research setting; thus, an ideographic methodology is used in this research (Creswell, 2013). Moreover, as the focus of this research is to deliver the criticisms of socio-environmental activists to develop a new socio-environmental accountability model (KPA, 2017; LBH-Jakarta, 2015; LBH-Makassar, 2016; Sukoharsono, 2007; WALHI, 2015, 2016; Watchdoc, 2018), this research will use critical paradigm as a basis view and explanation.

The critical paradigm is chosen because it focuses lies on the intention to liberate people from all chains or oppression (structured and not structured) in their daily life. This view is based on an understanding that humans are free beings to determine the meaning of their life (nominalism). Thus, the methodology used in the critical paradigm study is grounded on the criticism to one domination case [policy or system] and then it presents an antithesis as a solution to the concept that oppresses the community (Burrell & Morgan, 1979).

In this research, it is believed that the current accountability system of the local governments tends to be dominated by the views and practices that focus on monetary achievements or just focus on the investment and economic growth, not considering the non-monetary matters such as environmental, social, and culture issues. Although, many socio-environmental impacts affect the society, there is a massive movement from socio-environmental NGOs in criticizing the accountability activities (KPA, 2017; LBH-Jakarta, 2015; LBH-Makassar, 2016; WALHI, 2015, 2016; Watchdoc, 2018), and the sustainable development goals have become the main basis of government activities as required by laws.

Therefore, the idea of emancipatory socio-environmental accounting framework of Saravanamuthu and Lehman (2013) will be used as a framework to bridge the socio-environmental activists' views, sustainable development goals, and the current accountability problems for criticizing the current government's accountability framework and delivering a new integrated accountability framework as an antithesis for solving the socio-environmental problems in Indonesia. Thus, the research setting is as follows (Fig. 2.2).

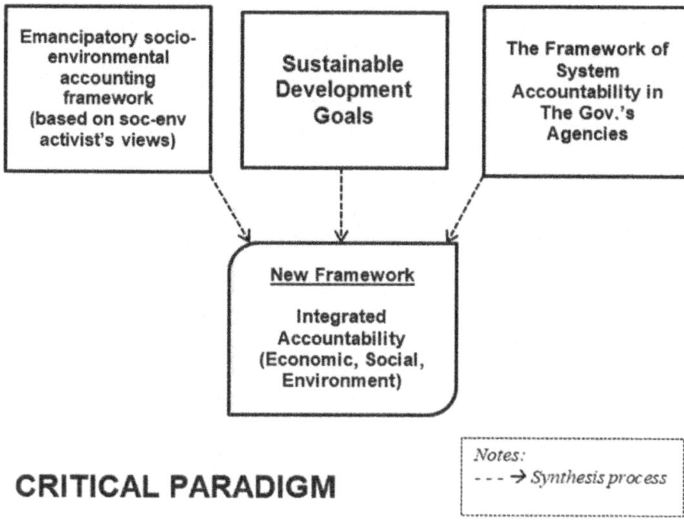

Fig. 2.2 Research setting for this book

References

AccountAbility. (2008). *AA1000 Accountability Principles Standard 2008*. Retrieved from www.accountability21.net.

Adams, C. A., & McNicholas, P. (2007). Making a difference: Sustainability reporting, accountability and organisational change. *Accounting, Auditing & Accountability Journal, 20*(3), 382–402.

Adeyeye, A. (2011). Universal standards in CSR: Are we prepared? *Corporate Governance: The International Journal of Business in Society, 11*(1), 107–119.

Adsera, A., Boix, C., & Payne, M. (2003). Are you being served? Political accountability and quality of government. *The Journal of Law, Economics, and Organization, 19*(2), 445–490.

Apple, M. W. (2005). Education, markets, and an audit culture. *Critical Quarterly, 47*(1–2), 11–29. https://doi.org/10.1111/j.0011-1562.2005.00611.x.

Archel, P., Husillos, J., Larrinaga, C., & Spence, C. (2009). Social disclosure, legitimacy theory and the role of the state. *Accounting, Auditing & Accountability Journal, 22*(8), 1284–1307. https://doi.org/10.1108/09513570910999319.

Aritonang, D. (1999). *Runtuhnya rezim daripada Soeharto: rekaman perjuangan mahasiswa Indonesia 1998*. Pustaka Hidayah.

Ball, A. (2004). A sustainability accounting project for the UK local government sector? Testing the social theory mapping process and locating a frame of reference. *Critical Perspectives on Accounting, 15*(4), 1009–1035.

Ball, A. (2005). Environmental accounting and change in UK local government. *Accounting, Auditing & Accountability Journal, 18*(3), 346–373. https://doi.org/10.1108/09513570510600738.

Ball, A. (2007). Environmental accounting as workplace activism. *Critical Perspectives on Accounting, 18*(7), 759–778.

Benmelech, E., & Moskowitz, T. J. (2010). The political economy of financial regulation: Evidence from US state usury laws in the 19th century. *The Journal of Finance, 65*(3), 1029–1073.

Boyne, G., & Law, J. (1991). Accountability and local authority annual reports: The case of Welsh district councils. *Financial Accountability & Management, 7*(3), 179–194.
Burrell, G., & Morgan, G. (1979). *Sociological paradigms and organisational analysis: Elements of the sociology of corporate life*. London: Heinemann.
Cameron, W. (2004). Public accountability: Effectiveness, equity, ethics. *Australian Journal of Public Administration, 63*(4), 59–67.
Cole, D. C., & Slade, B. F. (1998). Why has Indonesia's financial crisis been so bad? *Bulletin of Indonesian Economic Studies, 34*(2), 61–66. https://doi.org/10.1080/00074919812331337330.
Creswell, J. W. (2013). *Research design: Qualitative, quantitative, and mixed methods approaches*. Sage.
Davis, A. K., Guenther, D. A., Krull, L. K., & Williams, B. M. (2013). *Taxes and corporate accountability reporting: Is paying taxes viewed as socially responsible*. Lundquist College of Business Working Paper, University of Oregon.
Deegan, C. (2002). Introduction: The legitimising effect of social and environmental disclosures—A theoretical foundation. *Accounting, Auditing & Accountability Journal, 15*(3), 282–311.
Deegan, C. (2007). *Financial accounting theory*. Australia: McGraw-Hill Irwin.
Diefenbach, T. (2009). New public management in public sector organizations: The Dark sides of managerialistic "Enlightenment." *Public Administration, 87*(4), 892–909. https://doi.org/10.1111/j.1467-9299.2009.01766.x.
Djamhuri, A. (2009). *A case study of governmental accounting and budgeting Reform at local authority in indonesia: An institutionalist perspective*. USM.
Dowling, J., & Pfeffer, J. (1975). Organizational legitimacy: Social values and organizational behavior. *Pacific Sociological Review*, 122–136.
Dwivedi, O. P. (1989). Public service responsibility and accountability. In J. G. Jabbra & O. P. Dwivedi (Eds.),*Public service accountability: A comparative perspective*. Connecticut: Kumarian Press.
Dykstra, C. A. (1939). The quest for responsibility. *American Political Science Review, 33*(1), 1–25.
Eckardt, S. (2008). Political accountability, fiscal conditions and local government performance—cross-sectional evidence from Indonesia. *Public Administration and Development, 28*(1), 1–17. https://doi.org/10.1002/pad.475.
Elkington, J. (1998). Accounting for the triple bottom line. *Measuring Business Excellence, 2*(3), 18–22.
Elliott, J. A. (2012). *An introduction to sustainable development*. Routledge.
Ezzamel, M. (1997). Accounting, control and accountability: Preliminary evidence from ancient Egypt. *Critical Perspectives on Accounting, 8*(6), 563–601. https://doi.org/10.1006/cpac.1997.0123.
Fukuyama, F. (2006). *The end of history and the last man*. New York: Simon and Schuster.
Fukuyama, F. (2014). *Political order and political decay: From the industrial revolution to the globalization of democracy*. New York: Farrar, Straus and Giroux.
Gray, R. (2001). Thirty years of social accounting, reporting and auditing: What (if anything) have we learnt? *Business Ethics: A European Review, 10*(1), 9–15. https://doi.org/10.1111/1467-8608.00207.
Gray, R. (2002). The social accounting project and Accounting Organizations and Society Privileging engagement, imaginings, new accountings and pragmatism over critique? *Accounting, Organizations and Society, 27*(7), 687–708. Retrieved from https://www.sciencedirect.com/science/article/pii/S0361368200000039.
Gray, R, Owen, D., & Adams, C. (1996). *Accounting & accountability: Changes and challenges in corporate social and environmental reporting*. Retrieved from https://eprints.gla.ac.uk/95486/.
Gray, R., Bebbington, J., & Collison, D. (2006). NGOs, civil society and accountability: Making the people accountable to capital. *Accounting, Auditing & Accountability Journal, 19*(3), 319–348.
GRI. (2002). *Sustainability reporting guidelines*. Boston: Global Reporting Initiative.
GRI. (2016). *Consolidated set of GRI Sustainability Reporting Standards* (2016th ed.). Amsterdam: Global Reporting Initiative.

Grindle, M. S. (2004). Good enough governance: Poverty reduction and reform in developing countries. *Governance, 17*(4), 525–548.

Gunderson, L. H., Holling, C. S. (eds.). (1995). *Barriers and bridges to the renewal of ecosystems and institutions*. New York: Colombia Press.

Guthrie, & Parker. (1989). Corporate social responsibility practice: a rebuttal of legitimacy theory. *Accounting and Business Research, 19*(1), 343–352.

Hadi, S., Daeng, S., Afrimadona, Darmastuti, S., Pratiwi, E., & Nataprawira, I. (2012). *Kudeta Putih: Reformasi and Pelembagaan Kepentingan Asing dalam Ekonomi Indonesia*. Jakarta: Bentang Pustaka.

Hadiz, V. R. (2004). Decentralization and democracy in Indonesia: A critique of neo-institutionalist perspectives. *Development and Change, 35*(4), 697–718.

Harun, H., An, Y., & Kahar, A. (2013). Implementation and challenges of introducing NPM and accrual accounting in Indonesian local government. *Public Money and Management, 33*(5), 383–388. https://doi.org/10.1080/09540962.2013.817131.

Henisz, W. J., & Zelner, B. A. (2005). Legitimacy, interest group pressures, and change in emergent institutions: The case of foreign investors and host country governments. *Academy of Management Review, 30*(2), 361–382.

Hill, H., & Shiraishi, T. (2007). Indonesia after the Asian Crisis. *Asian Economic Policy Review, 2*(1), 123–141. https://doi.org/10.1111/j.1748-3131.2007.00058.x.

Hockings, M., Cook, C. N., Carter, R. W., & James, R. (2009). Accountability, reporting, or management improvement? Development of a State of the Parks assessment system in New South Wales, Australia. *Environmental Management, 43*(6), 1013–1025.

Islam, M., & Deegan, C. (2010). Media pressure and corporate disclosure of social responsibility performance: A case study of two global clothing and sports retail companies. *Accounting Business Research, 40*(2), 131–148.

Jennifer Ho, L., & Taylor, M. E. (2007). An empirical analysis of triple bottom-line reporting and its determinants: Evidence from the United States and Japan. *Journal of International Financial Management & Accounting, 18*(2), 123–150.

Jindrichovska, I., & Purcarea, I. (2011). CSR and environmental reporting in the Czech Republic and Romania: Country comparison of rules and practices. *Accounting and Management Information Systems, 10*(2), 202.

Joseph, C., Pilcher, R., & Taplin, R. (2014). Malaysian local government internet sustainability reporting. *Pacific Accounting Review, 26*(1/2), 75–93. https://doi.org/10.1108/PAR-07-2013-0071.

Julianto, E. (2010). *Dasar Pertimbangan dan Proses Perumusan Opini dalam Pemeriksaan atas Laporan Keuangan Daerah* . Retrieved from https://www.academia.edu/7113796/Dasar_Pertimbangan_Opini_Audit

KNKG. (2008). *Pedoman Umum Good Public Governance Indonesia*. Jakarta: Komite Nasional Kebijakan Governance (KNKG).

Kompas. (2017). *Fitra: Jangan Jadikan WTP Bahan Pencitraan Pemerintah*. Retrieved December 17, 2017, from https://nasional.kompas.com/read/2017/05/23/14022421/fitra.jangan.jadikan.wtp.bahan.pencitraan.pemerintah.

Koppell, J. G. S. (2008). Global governance organizations: Legitimacy and authority in conflict. *Journal of Public Administration Research and Theory, 18*(2), 177–203.

KPA. (2017). KPA Launching Catatan Akhir Tahun 2016. Retrieved April 17, 2017, from https://www.kpa.or.id/news/blog/kpa-launching-catatan-akhir-tahun-2016/.

Kristiansen, S., Dwiyanto, A., Pramusinto, A., & Putranto, E. A. (2009). Public sector reforms and financial transparency: Experiences from Indonesian districts. *Contemporary Southeast Asia: A Journal of International and Strategic Affairs, 31*(1), 64–87. https://doi.org/10.1355/CS31-1C.

Lanis, R., & Richardson, G. (2012). Corporate social responsibility and tax aggressiveness: An emperical analysis. *Accounting Public Policy, 31*(1), 86–108.

Lanis, R., & Richardson, G. (2013). Corporate social responsibility and tax aggresiveness: A test of legitimacy theory. *Accounting, Auditing & Accountability Journal, 26*(1), 75–100.

Lapsley, I. (2009). New public management: The cruellest invention of the human spirit? *Abacus, 45*(1), 1–21. https://doi.org/10.1111/j.1467-6281.2009.00275.x.
LBH-Jakarta. (2015). *Atas Nama Pembangunan: Laporan Penggusuran Paksa di Jakarta Tahun 2015* (Vol. 2017). Retrieved from https://www.bantuanhukum.or.id/web/atas-nama-pembangunan-laporan-penggusuran-paksa-di-jakarta-tahun-201.
LBH-Makassar. (2016). Siaran Pers Advokasi Masyarakat. Retrieved from https://lbhmakassar.org/siaran-pers/.
Macintosh, N. (2002). *Accounting, accountants and accountability: Poststructuralist positions* (2007th ed.). London: Routledge.
MacLullich, K. K. (2004). Accounting, love and justice. *Accounting Auditing and Accountability Journal, 17*(3), 327–360.
Maesschalck, J. (2004). The impact of new public management reforms on public servants' ethics: Towards a theory. *Public Administration, 82*(2), 465–489. https://doi.org/10.1111/j.0033-3298.2004.00403.x.
McLeod, R. H. (2000). Soeharto's Indonesia: A better class of corruption. *Agenda: A Journal of Policy Analysis and Reform*, 99–112.
Mebratu, D. (1998). Sustainability and sustainable development: Historical and conceptual review. *Environmental Impact Assessment Review, 18*(6), 493–520.
Merriam-Webster. (2018). *Definition of accountability*. Retrieved April 10, 2018, from https://www.merriam-webster.com/dictionary/accountability.
metrotvnews.com. (2017). *Pemerintah Cabut 23 Peraturan Menteri Penghambat Investasi*. Retrieved from https://news.metrotvnews.com/read/2017/04/04/681278/pemerintah-cabut-23-peraturan-menteri-penghambat-invest.
Mulawarman, A. D. (2016). *2024 Hijrah Untuk Negeri: Kehancuran atau Kebangkitan Indonesia dalam Ayunan Peradaban*. Jakarta: Yayasan Rumah Peneleh.
Mulgan, R. (2000). "Accountability": An ever expanding Concept? *Public Administraton, 78*(3), 555–573. https://doi.org/10.1111/1467-9299.00218.
Niemann, L., & Hoppe, T. (2018). Sustainability reporting by local governments: A magic tool? Lessons on use and usefulness from European pioneers. *Public Management Review, 20*(1), 201–223. https://doi.org/10.1080/14719037.2017.1293149.
OECD. (2018). *Accountability (in management theory)*. Retrieved May 11, 2018, from Glossary of Statistical Term-OECD website: https://stats.oecd.org/glossary/detail.asp?ID=4757.
Parker, L. (2005). Social and environmental accountability research: A view from the commentary box. *Accounting, Auditing & Accountability Journal, 18*(6), 842–860.
Patten, D. M. (1991). Exposure, legitimacy, and social disclosure. *Journal of Accounting and Public Policy, 10*(4), 297–308.
Qian, W., Burritt, R., & Monroe, G. (2011). Environmental management accounting in local government. *Accounting, Auditing & Accountability Journal, 24*(1), 93–128. https://doi.org/10.1108/09513571111098072.
Radaelli, C. M. (2005). Diffusion without convergence: How political context shapes the adoption of regulatory impact assessment. *Journal of European Public Policy, 12*(5), 924–943.
Rajiani, I. (2011). A cultural ecology of new public management in Indonesia. *Journal of Administrative Science, 8*(1), 17–31.
Ramanna, K. (2013). A framework for research on corporate accountability reporting. *Accounting Horizons, 27*(2), 409–432.
Ramli, R., & Nuryadin, P. (2007). Ten years after: Impact of monetarist and neoliberal solutions in Indonesia. *Economic and Political Weekly*, 79–88.
Rasyid, M. R. (2003). Regional autonomy and local politics in Indonesia. *Local Power and Politics in Indonesia*, 63–72.
RMOL.ID. (2015). Demo senin tuntut lestarikan hutan, cabut izin perusahaan pembakar lahan. Retrieved from https://politik.rmol.id/read/2015/09/27/218798/demo-senin-tuntut-lestarikan-hutan-cabut-izin-perusahaan-pembakar-lahan.

Robison, R., & Rosser, A. (1998). Contesting reform: Indonesia's new order and the IMF. *World Development, 26*(8), 1593–1609.

Sanusi, A., Butler, B., & Rola-rubzen, M. F. (2013). Local politics and business in Indonesia: A non-market strategy approach. In *Provincial Governance Strengthening Programme Supported by UNDP Indonesia* (pp. 1–20). Unika Atmajaya, Jakarta: Unika Atmajaya.

Saravanamuthu, K. (2004). What is measured counts: Harmonized corporate reporting and sustainable economic development. *Critical Perspectives on Accounting, 15*(3), 295–302. https://doi.org/10.1016/S1045-2354(03)00063-7.

Saravanamuthu, K. (2009). A journey of socialising the risks associated with global warming: A Gandhian insight into Schumacher's total accounting and accountability. In *Extending Schumacher's Concept of Total Accounting and Accountability into the 21st Century* (pp. 121–189). https://doi.org/10.1108/S1041-7060(2009)0000014008.

Saravanamuthu, K., & Lehman, C. (2013). Enhancing stakeholder interaction through environmental risk accounts. *Critical Perspectives on Accounting, 24*(6), 410–437. https://doi.org/10.1016/j.cpa.2013.02.002.

Schedler, A., Diamond, L. J., & Plattner, M. F. (1999). *The self-restraining state: power and accountability in new democracies.* Lynne Rienner Publishers.

Seidman, G. I. (2004). The origins of accountability: Everything I know about the sovereign's immunity, I learned from King Henry III. *Louis ULJ, 49*, 393. Retrieved from https://heinonline.org/HOL/LandingPage?handle=hein.journals/stlulj49&div=22&id=&page=.

Shahib, H. M. (2016). Financial Regulation Violations and Corporate social responsibility disclosure: A test of legitimacy theory in Indonesia Stock Exchange. In *Proceeding of International Conference on Accounting, Management, Economics and Social Sciences (ICAMESS) 2016* (pp. 815–824), Millenium Hotel, Jakarta, 30 April 2016, ISBN: 978-602-73705-0-0. Retrieved from https://papers.ssrn.com/sol3/papers.cfm?abstract_id=2891588.

Shahib, H. M., & Irwandi, S. A. (2016). Violation regulation of financial services authority (FSA), financial performance, and corporate social responsibility disclosure. *Journal of Economics, Business, and Accountancy Ventura, 19*(1), 141–154. https://doi.org/10.14414/jebav.v19i1.533.

Shahib, H. M., & Risky, F. R. (2017). Accountability in the Internet Era: A lesson from local governments in Indonesia. *Hasanuddin Economics and Business Review, 1*(1), 57–74.

Sindonews.com. (2016). Pemerintah Beri Keringanan Amdal Perusahaan di Kawasan Industri (15 Maret 2016, Ed.). Retrieved from https://ekbis.sindonews.com/read/1093165/34/pemerintah-beri-keringanan-amdal-perusahaan-di-kawasan-industri-1458039020.

Stanton, P., & Stanton, J. (2002). Corporate annual reports: Research perspectives used. *Accounting, Auditing & Accountability Journal, 15*(4), 478–500.

Stiglitz, J. E., Sen, A., & Fitossi, J.-P. (2009). *Commission on the measurement of economic performance and social progress report-2009.* Paris: The Commission on The Measurement of Economic Performance and Social Progress-France.

Sugandhy, A., & Hakim, R. (2007). *Prinsip dasar kebijakan pembangunan berkelanjutan berwawasan lingkungan.* Bumi Aksara.

Sukoharsono, E. G. (2005). Environmental accounting: A map of environmental regulations and policies of the Indonesian Central Government. *The International Journal of Accounting and Business Society, 13*(2), 1–22.

Sukoharsono, E. G. (2007). Green accounting in Indonesia: Accountability and environmental issues. *The International Journal of Accounting and Business Society, 15*(1), 21–60.

Tallontire, A. (2007). CSR and regulation: Towards a framework for understanding private standards initiatives in the agri-food chain. *Third World Quarterly, 28*(4), 775–791.

Timur, T. (2015). Pemerintahan Jokowi harus lindungi ekosistem karst yang terancam industri pertambangan. Retrieved from https://www.tribunnews.com/regional/2016/01/11/pemerintahan-jokowi-harus-lindungi-ekosistem-karst-yang-terancam-industri-pertambangan.

Tinker, A., & Gray, R. (2003). Beyond a critique of pure reason: From policy to politics and praxis in environmental and social research. *Accounting, Auditing, and Accountability Journal, 16*(5), 727–761.

Tschopp, D. J. (2005). Corporate social responsibility: A comparison between the United States and the European Union. *Corporate Social Responsibility and Environmental Management, 12*(1), 55–59.

UNDP. (2018). *Sustainable development goals*. Retrieved May 15, 2018, from United Nations Development Programme website: https://www.id.undp.org/content/indonesia/en/home/sustainable-development-goals.html.

Vogel, D. (2010). The private regulation of global corporate conduct: Achievements and limitations. *Business & Society, 49*(1), 68–87.

WALHI. (2015). *Tinjauan Lingkungan Hidup 2015 WALHI*. Retrieved from https://walhi.or.id/wp-content/uploads/2016/12/OutLook2015_W.pdf.

WALHI. (2016). *Tinjauan Lingkungan Hidup WALHI 2016: Keharusan Pembenahan Struktural untuk Perbaikan Tata Kelola*. Retrieved from Wahana Lingkungan Hidup Indonesia website: https://www.walhi.or.id/2016/12/22/outlook-2016-tujuan-lingkungan-hidup/.

Waluyo, W. (2004). Lingkungan dan Pembangunan Berkelanjutan. *Yustisia, 65*, 890–896.

Watchdoc. (2018). *Kumpulan film dokumenter kerusakan lingkungan di Indonesia*. Retrieved February 14, 2018, from Youtube website: https://www.youtube.com/user/watchd0c/featured.

Weber, J., & Wasieleski, D. M. (2013). Corporate ethics and compliance programs: A report, analysis and critique. *Journal of Business Ethics, 112*(4), 609–626.

Williams, C. (2006). *Leadership accountability in a globalizing world*. London: Springer.

Williams, B., Wilmshurst, T., & Clift, R. (2011). Sustainability reporting by local government in Australia: Current and future prospects. *Accounting Forum, 35*, 176–186.

World-Bank. (2012). *Doing business in Indonesia 2012*. Washington, DC: World-Bank.

Chapter 3
Research Methodology

From the way people think we can consider their philosophy, also, from their philosophy, we may know what method they use to come to that philosophy.
—Tan Malaka (Indonesian's National Hero and intellectual)

The previous chapters have discussed the reality portrait of Indonesian local government's accountability framework, critiques of social-environmental activists on it, and the opportunity to do an emancipatory socio-environmental accounting research that can develop an integrated government accountability system. Now, this section is to present a research methodology discussion that will lead to the new antithesis or better form of government accountability framework that integrates the economic-socio-environmental issues, following the aspirations of Indonesian socio-environmental activists, society, as well as supporting the achievement of sustainable development goals that became a worldwide concern today.

3.1 Critical Paradigm

By referring to the literature, what is the right paradigm to be used as a basis to criticize or to find a solution to a structure or condition that tends to oppress or create many problems to people? The critical paradigm is generally be the answer (Burrell & Morgan, 1979; Chua, 1986; Creswell, 2013; Kamayanti, 2016). The paradigm departs from the philosophical tradition that ontologically believes that human beings give meaning to their life liberally (nominalism), and human nature is determined freely (voluntarism). This view then ultimately leads to the use of an anti-positivism epistemology, which believes the truth and liberation can be revealed by profoundly understand the people who involved in one dominating case. The central thinkers of critical paradigm is Karl Marx, leftist thinkers (Marxists) who continuously refer to Marx's ideas, e.g. Trotsky, Antonio Gramsci, Louis Althusser; and Frankfurt School (critical theory) thinkers, e.g. Jurgen Habermas, Theodore W Adorno, Herbert Marcuse, and Erick Fromm (Burrell & Morgan, 1979; Chua, 1986; Ritzer & Stepnisky, 2017).

H. M. Shahib, *Towards the Local Government's Integrated Accountability Framework*,
https://doi.org/10.1007/978-981-33-6617-6_3

Furthermore, axiologically, there are two common approaches to the critical paradigm manifestation in practical actions to change oppressions or unfair situations, namely radical structuralism and radical humanism approach. Radical structuralism approach focuses on changing oppressive structures directly. On the other hand, radical humanism approach gives its main attention on enlightening individual consciousness to lead into the community awareness, to bring them a radical liberating change from a domination or unfair situation (Burrell & Morgan, 1979; Kaltenbrunner, Newman, & Painceira, 2012; Kamayanti, 2017; Mulawarman, 2010; Ryer, 1999; Toft, 2015).

Moreover, in order to map the differences between the two main radical approaches above, people can differentiate them by their particular actions in liberating the oppressed society. For instance, radical structuralism actions tend to focus on doing the right critic movement to the government or economic structure and policy that perceived dominate the society. The terms such as demonstration, resistance, class struggle, the means of production takeover, system reconstruction, and revolution are general terms used by radical structuralism approach (Burrell & Morgan, 1979; Gramsci, 1994; Ritzer & Stepnisky, 2017). On the other hand, radical humanism approach is known by its awareness movements to the people. Educating the individual to be conscious that there is unfair domination in their life by using counter-discourse, actively doing solidarity actions, consensus-based dialog, direct action, and doing self-management are general characteristics of radical humanism approach (Burrell & Morgan, 1979; Cleaver, 2017; Ritzer & Stepnisky, 2017). These actions can be easily found in anarchist communities today (Mayer, 2008).

Apart from the discussion of critical paradigm and its variances in the practical field, there is also extensive discussion related to the use of a critical paradigm as a basis of research methodology in the socio-economic research field. In general, socio-economic studies applying the critical paradigm also focus on discussion, criticize, and seek to find a solution related to the oppressed and unfair situation that a society is unable to resolve. For instance, the studies related to the severe economic policy (Llorente, 2006; Springer, 2012), oppressed government's system (Cleaver, 1998), and culture hegemony (Krishnaswamy, 2002; Ritzer & Jurgenson, 2010). In the same way, several studies have used critical paradigm as a research methodology basis to criticize and attempt to discover a solution related to dominating and inequitable issues in accounting. Accounting's policy, standard, system, and culture (Macintosh, Hopper, & Figure, 2005; Ryer, 1999; Tinker & Gray, 2003).

In line with the critical paradigm discussion above, there is a correlation between this research's objective and the rationale for critical paradigm to become this research's basis view and explanation. As in this research case, it is believed that the current accountability system in local governments have a tendency to be dominated by the views and practices that focus on monetary achievement or only on the investment and economic growth, not in the non-monetary matters such as the environmental, social, and cultural issues which by far, are worse. Therefore, it is believed, the critical paradigm is possible to be the basis to explain, criticize, and deliver an antithesis of the current accountability system. Especially, on the attempt to resolve socio-environmental accountability problems in Indonesia that have become the primary concern of socio-environmental NGOs today in their critics towards government's accountability activities (KPA, 2017; LBH-Jakarta, 2015; LBH-Makassar, 2016; WALHI, 2015, 2016; Watchdoc, 2018).

However, a question is then arising, what will the appropriate critical thought be used as an explanation and methodological basis for this research theme? It is because the critical paradigm as a school of thought is prosperous of ideas from many scholars with different backgrounds and socio-historical contexts. To answer this question, please see the following explanation in Sects. 3.2 and 3.3.

3.2 Contextuality and Locality of Critical Paradigm

The critical paradigm discourse has uniqueness, specifically on its thought and methodology discourses. It is because its theories and methodologies are incline only to have similarity on the ontological side, that believes humans form their social reality, and apply the historical-material dialectic epistemology to free the oppressed people. Both critical theories and methodologies are likely to be diverse from the axiological or implementation side.

These situations are understandable because the critical intellectuals lived in different times and environment with different challenges that would require different solutions (axiology). Mainly, the critical scholars also widely believe in the dictum of historical-dialectic materialism that requires changes in addressing social orders (Burrell & Morgan, 1979; Ritzer & Stepnisky, 2017; Silverman, 1985). Thus, contextual and locality of the critical paradigm reflected in different movements and ideas among critical thinkers around the world become inevitable (Ritzer & Stepnisky, 2017).

As can be seen from the thoughts of thinkers who claim themselves as critical thinkers, especially critical thinkers after Marx, there are so many different ideas among them although they lived in the same period or even in the same country. For example, Trotsky (1879–1940) and Stalin (1878–1953) who claimed they were Marxists and lived together in the same era and country. However, they had a significant difference on their views over the Marxist theory's implementation (axiology). Trotsky with his theory of permanent revolution believed the working class, especially those in the weak economy, must ride the bourgeois democratic revolution first, before creating the working class revolution, and the cooperation among countries' working classes can facilitate the socialist revolution to take place sooner (Trotsky, 2010).

In contrast, Stalin believed in the working class, and peasants within one State are sufficient to carry out a revolution, and in the end, it will be up as a Worker's State that will control the state in an autocratic way for the benefit of the working class (Stalin, 1978). The different perspective also comes from the other Marxist and critical thinker from Italy, Gramsci (1891–1937). He believed in the axiology that besides the class struggle, it is essential for the working people to understand and destroy the hegemonic work of capitalism in the form of ideology and culture. It is due to the ideological and capital hegemony that are interweaved to continue and strengthen the era of capitalism (Gramsci, 1994; Nezar & Andi, 2009).

Furthermore, the different views of Marxist axiology and critical theory also come from a thinker of the Frankfurt school of thought, Habermas (1929–present) who believes a change in the oppressive social structure of the modern era can be presented with two forces (approaches). First, technical management based on the rational decision in the production work area. Government intervention is an example. Second, in the area of social interaction, people need to make a mutual understanding or consensus based on public's interest that put aside the ideological difference for the common interest (Habermas, 1984). Similarly, Harry Cleaver, a modern critical thinker from the United States, believes in the era of modern capitalism, in praxis (axiology) field, the working class needs the development of horizontal worker's organizations, consensus-based decision-making and anti-hierarchical organization as the solution to change the construct of capitalism in this twenty-first century society. This thought is popularly known as the autonomist Marxism (Cleaver, 1998, 2017).

According to the discussion above, it then becomes interesting to see the application of various ideas of critical scholars in the socio-economic research field, specifically in accounting studies. It can be seen that there is so rich difference reading from critical paradigms and critical methodologies on accounting phenomena as the research topic. For example, Ryer (1999) used Marx's theory of surplus value to criticize the FASB's conceptual framework; Lehman (2001) used the Habermas theory, particularly the public sphere, to find a consensus of scientific discussion on corporate socio-environmental accounting; and a study by Spence (2009) used Gramsci's idea of hegemony in criticizing social accounting practices. From these articles, it can be seen that the variety of readings is due to the use of critical paradigms in critical methodologies heavily depend on the coherence of ontology-epistemologies, context of a case, and scholar's preliminary assumptions about the research setting.

However, there are also other facts about the use of critical scholars' ideas, specifically the western critical-scholars' thoughts, as a methodology and an analysis tool in accounting studies. It could be said; there are smooth hegemony-movement of western's critical view upon accounting discourse around the world (Alawattage & Wickramasinghe, 2008; Gallhofer et al., 2000; Macintosh et al., 2005), one thing that strongly opposes the principle of the critical paradigm development itself. Moreover, it also could be said that there are several blind spots in the mainstream of western's critical thoughts specifically related to their worldview.

In general, this claim has three reasons. Firstly, the use of critical thoughts in accounting research is likely to be based on one fundamental assumption. There is a coherence situation between the research background and western critical theory to be used as a methodology or analysis tool(Laughlin, 1987). Accounting scholars are likely to set aside to consider that the construction of a western critical idea is very closely related to the social background in which a critical thinker lived (da Fonseca, 2010; Escobar, 1998; Finnemore, 1996; Gallhofer et al., 2000; Said, 2012). As widely known, the mainstream critical thinkers are western people. Thus, this fact will bring a consequence to the social structure and practice in a research setting.

The study site's social circumstances could gradually absorb the "western taste".[1] In the end, it is not suitable for the local characteristics existence (Bhabha, 1984).

Secondly, the blind spot of the mainstream western's critical theories is characterized by how they incline to see the social reality in binary-opposition perspective, in structuralism and non-structuralism approaches (Kamayanti, 2016; Mulawarman, 2010; Powell, 1998; Ritzer & Stepnisky, 2017). Western's critical thoughts generally have not seen both approaches could influence each other. Although there are already several western thinkers who begin to speak it out, for instance, Giddens (1984) and Foucault (1972). Thirdly, another weakness of the western's critical ideas, they tend to exclude local intelligence and wisdom in order to explain the local problems (Gallhofer et al., 2000). Analogically, it is like giving glasses to a blind person. "The glasses" does not touch the core of the problem.

Based on the reasons above and ideas supporting to review the use of mainstream western [critical] theories in a different and unique context (Duarte, da Fonseca, & da Fonseca, 2010; Escobar, 1998; Gallhofer et al., 2000; Kamayanti, 2017), this research will use a critical theory that is close to the context of Indonesian social life, as a basis to form ontology-epistemology-axiology-locality coherence in this study. As a result, the choice then falls into the critical thought of Tan Malaka from his book that entitled "Parliament or Soviet" (Malaka, 1921).

The use of TanMalaka's critical idea seems to offer more promise. It is because his thought covered up the ontology-epistemology-axiology-locality coherence. He was a productive critical philosopher with many books,[2] and more importantly, he was an independence warrior of Indonesia.[3] Thus, he was a real Indonesian thinker who knows the basic history and characteristic of Indonesia, as a nation and as a state (Poeze, 2008).

3.3 Tan Malaka's Critical Thought as a Methodology

To make it more clearly, the reason why I decided to choose Tan Malaka's critical thought to be the basis of this research methodology could be found on the following explanation. Philosophically, this research's problems can be seen ontologically as the domination of monetary-oriented view and performances that based on New Public Management (NPM) paradigm in the local government accountability practices (Diefenbach, 2009; Djamhuri, 2009; Rajiani, 2011). On the axiological level, this situation makes values and aspirations of other groups are being ignored (specifically, by the socio-environmental activists), and societies become the victim of

[1] In social discourse studies, this process is called "mimicry", a process to resemble one condition. From a cultural perspective, a person and a group of people may unwittingly have carried out the mimicry process because they have accustomed to do an action with other people's styles (Bhabha, 1984).

[2] See his books on https://www.marxists.org/indonesia/archive/malaka/index.htm.

[3] Presidential Decree No. 53 of 1963 of the Republic of Indonesia of March 28, 1963, concerning the title of the national hero of Tan Malaka.

social and environmental problems emerged from the government's development and accountability policies (KPA, 2017; LBH-Jakarta, 2015; LBH-Makassar, 2016; WALHI, 2015, 2016).

As can be seen from the argument above, there are three key phrases namely the (1) **domination of perspective and practice in government's practices**, (2) **ignorance to values and aspirations of other groups, and (3) communities as the victims**. Interestingly, Indonesia had a tremendously influential critical thinker who studied these three points long time ago. He even studied it directly with concern to the governance of an independent State (Indonesia) which he fought with all of his soul. He was Ibrahim Datuk Tan Malaka (1897–1949), a writer, a critical philosopher, as well as an independence national hero of Indonesia (Poeze, 2008).[4]

Tan Malaka's views, related to the domination of a worldview, especially the power of capitalism ideology, ignorance to values and aspirations of other groups, lead to the presence of depraved government policies that loss the people, is brightly discussed in his book entitled "Parliament or Soviet?" (Malaka, 1921). In this book, Tan Malaka warned Indonesians long time ago that western model government's management style adoption does not guarantee it will be suitable and effective to be applied in Indonesian context (Malaka, 1921). As he said:

> Throughout our knowledge, it is enough to show the nature, origin, and institutions of the West, one of which we must choose if we have the freedom to govern our own country (Malaka, 1921: 42)...... All domestic institutions without nationality would undoubtedly corrupt the nation (Malaka, 1921:43).

Moreover, Malaka (1921) added that the situations can be worse. If the government's governance system is based on class inequality, for example, no representation of clerics, the poor, labors, and peasants in regulating and controlling the implementation of the government's governance. Undoubtedly, it can be assured that the government system will not be focused on interests of the entire community. It will only benefited a well-rich group (capitalists) that can influence the government through its capital powers (Malaka, 1921). As he said:

> If we rethink, members of the Council come from a high class that has no interference with the people, [they do not] feel the trouble of the Kromo (little people). Undoubtly to say that members of the Council are not people's representatives and we are not surprised if [their] regulations opposing the needs of the people (Malaka, 1921: 6).
>
> Let us not be deceived if we hear that in a Parliament there are also the proletarian's deputies. They exist, but they have very little influence over the government of the country (Malaka, 1921: 28).
>
> The government is categorized over the "lawmaker" (Parliament) and "the one who runs the law" (minister) ...... the minister is full of bureaus [the bodies that establish, propose and enforce the law], then "lawmakers" and the one who run the "laws" are in fact ministers, too. Moreover, [it is due to] the interconnection of ministerial bureaus with banks and so on (the capitalist that have interests), then a country that has the Parliament [even based on the people's democratic voice], is willing to charge taxes for the land and sea military actions, robbed the land or people's treasures, and willing to allow the war, simply, the government

[4]For many scholars, Tan Malaka is categorized as an Indonesian Leftist and Socialist (see: Poeze (2008) and https://www.marxists.org/indonesia/archive/malaka/index.htm).

> is always enthusiastic to follow the Capitalists' will, which are never satisfied for profit (Malaka, 1921: 30).

Furthermore, Tan Malaka believed that the antithesis of the domination of world-views and practices, ignorance to values and aspirations of other groups, and people become the victims, is through people's political organizations (Tan Malaka called it as "committees") which fight for the people's interests in each economic and industry sectors. In the end, it is hoped that regulations based on the societies' interests will be presented and become the basis to oversee the government and industry performance for the sake of the people. As Malaka (1921) wrote:

> The two weapons of the workers (people), who will speak for their intentions in the future, are trade unions and [people] political organizations. The first, trade union's obligation to take up economic weapons, which can only be sharpened when it is entered politics that is the politics of the workers (people) themselves (Malaka, 1921:33).
>
> In every factory, mining, or spoor, the workers (people) themselves have held committees in which institutions that are now under their control. This (people) workers' committee is now the master of business; it checks each production of the factory, which is now being run no longer for the Capitalists who just know the profit, but for the working people [and society's interests] (Malaka, 1921: 34).

Based on Tan Malaka's thought discussion above, three main points can be linked to the current context, specifically about accommodating the view of social-environmental activists who expect the presence of a comprehensive model of government accountability. The main points are:

(1) Tan Malaka's view emphasizes the rejection of "Western" concept direct adoption in Indonesian government management system. In line with today's context, this condition also occurs in government accountability management systems that tend to adopt NPM models as a neo-liberalism derivative (Diefenbach, 2009; Djamhuri, 2009; Kristiansen, Dwiyanto, Pramusinto, & Putranto 2009; Mulawarman, 2016). Thus, the governments' accountability practice today tends to be profit-oriented, money based, and is unlikely to accommodate other references, such as local society's views and socio-environmental standards that can be found in sustainable development goals (UNDP, 2018). Therefore, there are a large number of socio-environmental activists groups then criticize it (KPA, 2017; LBH-Jakarta, 2015; LBH-Makassar, 2016; WALHI, 2015, 2016).

(2) Tan Malaka's view stresses the importance of representation of every class in the community (e.g., poor people, clerics, labors, farmers, employers, activists) to be given a place in the building of a government management system. Therefore, the government management system should be oriented to the public interests, which is also following the philosophical foundation of Indonesia (Pancasila) (Latif, 2011). Today, the contradictory situation takes place, as reflected in the presence of massive activist protests to encourage the governments to be more accountable in social and environmental issues, as a reaction to the adverse effect of the economic growth policy (LBH-Jakarta, 2015; WALHI, 2015, 2016).

(3) Tan Malaka's thought focuses on how social structures and agent (individual/society) can influence each other. Thus, the policy and regulations can

be based on societies' interests and be the basis to oversee the government and industry performance for the sake of the people.

Furthermore, when it is related to the discussion of the idea of emancipatory socio-environmental accounting framework of Saravanamuthu and Lehman (2013), there is always a possibility to make a new account based on the society's views that can be used as a new accounting tool to resolve a specific case. It is in harmony with the thought of Tan Malaka (1921), which focuses on the representation of people's aspirations in building the development of government management practices, especially those related to the socio-economic activities of Indonesians. Thus, it can be said that Tan Malaka's views can become a methodology basis of a critical-research attempts to elaborate emancipatory theme, specifically in Indonesian context. Furthermore, Tan Malaka's methodology is also aligned with what Sarantakos (2012) proposes that the methodology comesfrom the principles of a paradigm within a framework of acquiring knowledge and explaining it. For natural explanation, Tan Malaka's methodology can be described as follows (Fig. 3.1).

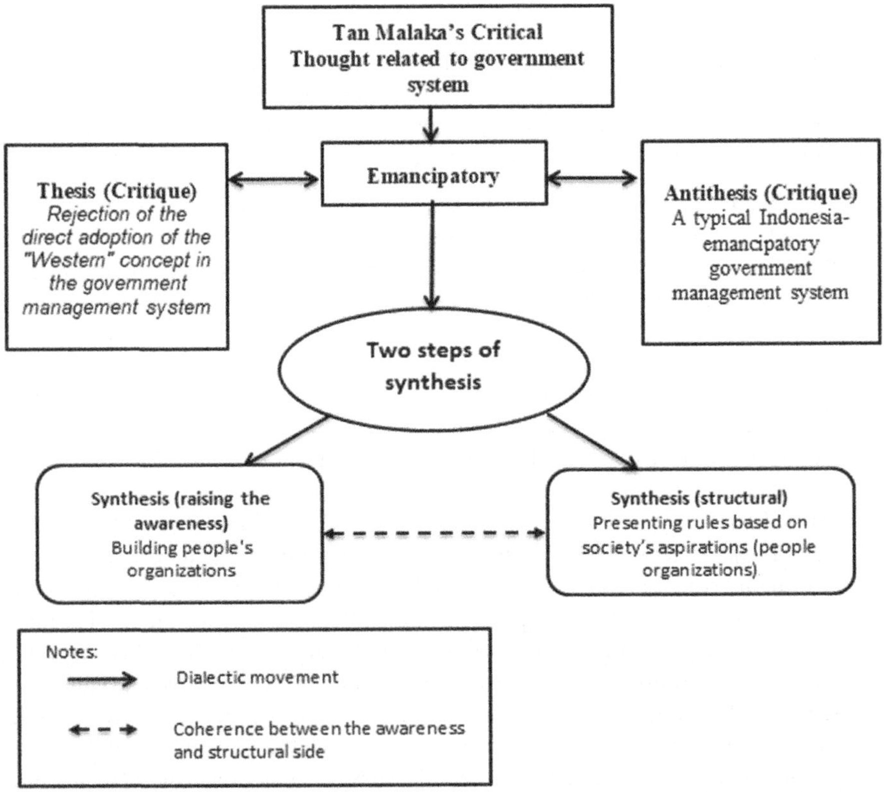

Fig. 3.1 The thought of Tan Malaka based on Tan Malaka's book that entitled "Parliament or Soviet? (1921)"

After discussing the philosophy of Tan ideas and their relations with socio-environmental activists critiques related to the performance of government accountability, the world's discourse on sustainable development goals, and emancipatory socio-environmental accounting framework of Saravanamuthu and Lehman (2013) that in line with Tan Malaka's emancipatory thought. Naturally, the description of this research's model to develop the integrated accountability framework, specifically for local governments in Indonesia, is in Fig. 3.2.

Based on the Tan Malaka's emancipatory methodological lens above, it can be realized that this research attempts to criticize and develop the system of government accountability model which currently tend to have NPM's characteristics such as tends to be profit-oriented, money-based, and are unlikely to accommodate other references (i.e. local society's views and socio-environmental standards). Moreover, this research also uses activists' views (critiques) on the government's socio-environmental accountability practices (KPA, 2017; LBH-Jakarta, 2015; LBH-Makassar, 2016; WALHI, 2015, 2016) and the central discourse related to the United Nation's sustainable development goals (SDGs). The results of the collection data then will be analyzed through synthesis-steps by using in-depth study, to develop a

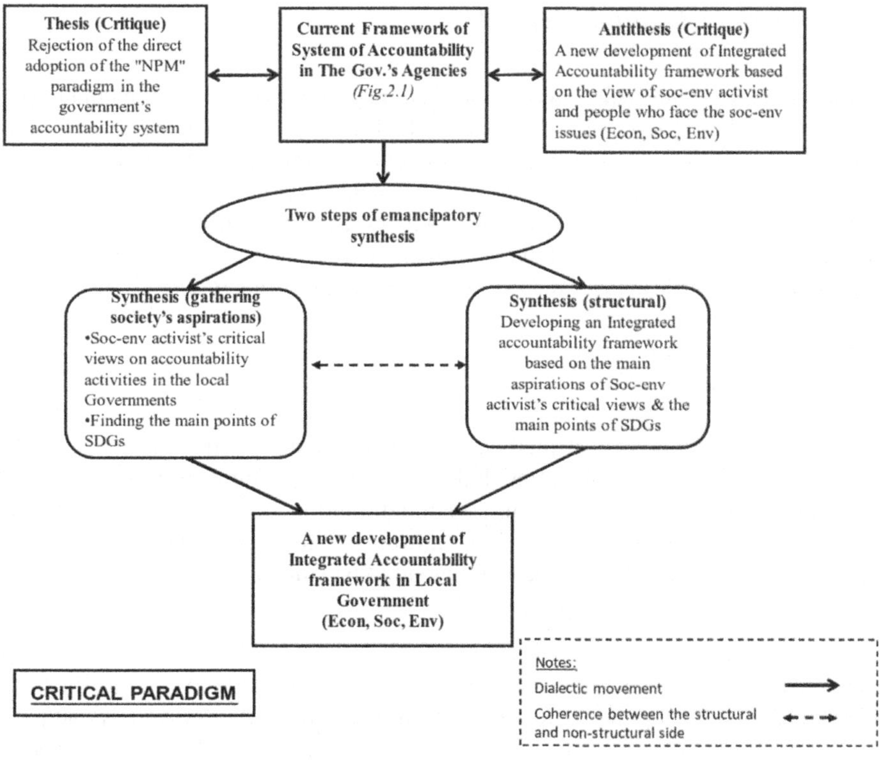

Fig. 3.2 Research model based on Tan Malaka's thought (Emancipatory framework for Indonesian context)

new comprehensive accountability framework for local governments. In particular, to achieve a balance of government accountability in the three main supporting dimensions of community life, namely economic, social and environmental dimensions (Diamond, 2013; Stiglitz, Sen, & Fitossi, 2009).

3.4 Research Design

To achieve a new accounting development, as this research wants, inevitably, methodological follow-up is required in the form of research method (Kamayanti, 2017). As this research appplies an emancipatory socio-environmental accounting concept by using Tan Malaka's emancipatory methodology to develop a new integrated accountability framework in the local governments. The coherence between the methodology and the method chosen is needed (Sarantakos, 2012). Therefore, the presentation of the research method of this research is as follows.

3.4.1 Data Collection Method

In order to collect data, this research uses the interview, documentation, and observation approaches. These data collection methods are considered appropriate for this study, as the methodology developed in this study requires an in-depth study related to the involvement of activists in doing advocacies over the social-environmental cases that have been missed by the government accountability in several Indonesian regions. Moreover, the following explanations will give more detail about the implementation of these three data collection methods in this research.

3.4.1.1 Interview

In line with the coherence of methodological reasoning of this research, the Tan Malaka's emancipatory values and linkages of the structural and non-structural synthesis are main basis of the semi-structured interview questions in this section. Thus, the interview questions to key informants, namely social-environmental activists and government employees, will tend to be similar.

1. *Socio-environmental activists*

This research puts the socio-environmental activists' aspiration to become the basis to evaluate and develop the socio-environmental accountability system as one form of the emancipatory socio-environmental accounting concept. Therefore, the socio-environmental activists of NGOs that have focused on doing advocacy activities

Table 3.1 The initial main locations and informants of this study[a]

No	Social-environmental cases	Cities/Districts	NGOs-Activists
1	Coastal reclamation	Jakarta	WALHI, RUJAK, KPA
		Makassar	WALHI
		Bali	WALHI
2	Environmental issues in the mine area	Palu-Donggala, Central Sulawesi	WALHI
		South Sulawesi	WALHI
		Rembang, Central Java	WALHI,Daulat Hijau
3	Eviction of people's homes for developmental reasons	Jakarta	RUJAK, KPA, LBH
		Makassar	LBH
		Kulonprogo, D.I Yogyakarta	Daulat Hijau

[a]WALHI: *Wahana Lingkungan Hidup Indonesia* (the Indonesian Forum for Environment); RUJAK *(Ruang Jakarta)*: a think act-tank that focus in advocating urban issues in Jakarta; KPA: *Konsorsium Pembaruan Agraria* (Agrarian reform consortium); Daulat Hijau: *Front Nahdliyin untuk Kedaulatan Sumber Daya Alam* (Nahdliyin Front for Sovereignty of Natural Resources); LBH: *Lembaga Bantuan Hukum* (Legal Aid Agency). The number of locations and informants have been adjusted in line with the development of the study

over the social-environmental issues in several Indonesian regions are the principal informants of this study.

At first, this study has a time limit. Consequently, it tends to focus on significant (having wide public attention) social-environmental cases in several cities and NGOs focusing on advocacy actions in those cases.[5] The initial data was as follows (Table 3.1).

However, there were several obstacles to cover all locations and informants above. One of them was the fact that there was no communications feedback from NGO activists and local people in a number of locations, and the great earthquake in Palu and Donggala (Central Sulawesi) in September 2018[6] with severe damage and difficulties to access these locations. Also, the limited time and funding. Consequently, this study was able only to cover the coastal reclamation in Makassar (south Sulawesi), as one main location to be studied deeply related to the socio-environmental activists' advocacy activities, government's accountability practices, and the affected people. Furthermore, for the remaining objects intended to be studied which cannot be covered, this research try to discuss them by using other sources such as the annual report of socio-environmental NGOs, news, and government's accountability documents related to those socio-environmental issues.

[5]This study is conducted in a range of 2017–2019 in line with the rise of advocacy on socio-environmental issues in Indonesia.

[6]See the news of a great earthquake in Palu (28 September 2018): https://www.bbc.com/news/world-asia-45683630.

Table 3.2 The summary of informants from NGOs in this study

Codes of informants	Social-environmental cases	Cities/Districts	Category of informants	Number of informants (FGD)	Time
WSS1-2	Coastal reclamation	Makassar, South Sulawesi	WALHI-South Sulawesi	2	3/12/18 (±90 min)
FN1-3			FORMASI NEGARA[a]	3 (FGD)	23/12/18 (±80 min)
WSS1-2	Environmental issues in the mining area	Galesong, South Sulawesi	WALHI-South Sulawesi	2	3/12/18 (±90 min)
FN1-3			FORMASI NEGARA	3 (FGD)	23/12/18 (±80 min)
WNE1		Maros and Pangkep, South Sulawesi	WALHI-National Executive, Jakarta	1	4/3/19 (±80 min)
		Rembang, Central Java			
KPA1[b]	Eviction of people's homes for development reasons (agrarian conflict)	All regions in Indonesia	KPA-National Executive, Jakarta	1	4/3/19 (±40 min)

[a] *Forum Masyarakat Pesisir-Nelayan Galesong Raya* (FORMASI-NEGARA) is a people forum of coastal communities and fishermen in Galesong district, Takalar Regency, South Sulawesi. The Galesong district is one of the coastal areas with social problems and environmental damage caused by Makassar reclamation project as Galesong's coastal area has been used as quarry for heap material in Makassar reclamation project

[b] Attended the KPA press conference regarding the evaluation of land reform policies (2014–2018) at the Indonesian Ombudsman's office in Jakarta on March 4, 2019

Finally, the interview process started with the director of WALHI in South Sulawesi.[7] Using snowball method, this study continued to select the other informants for the next step of interview activity. The following list was the main informants resulted from the snowball method.

The semi-structured interview questions based on Tan Malaka's emancipatory methodology values become the main guidance to interview the informants. The values of Tan Malaka's thoughts focus on three main points namely, (1) rejection to the direct adoption of western (foreign) styles in the government's management system (in this research case, it is NPM styles), (2) the importance of representation of every class and groups in the community to be given a place in the [process of] building a government management system, and (3) focus on how the social structures and agent (individual/society) can influence each other. Therefore, the following Table 3.2 shows the interview questions with substancial values of Tan Malaka's critical methodology as used in this research (Table 3.3).

[7] Director of WALHI-South Sulawesi (2018–2022) is Muhammad Al-Amin.

This extensive interview focuses to obtain primary information related to how activists see today's government's accountability activities and report and their views about what the governments should do in order to better accommodate the societies' views on the creation of accountability activities and report. The collection of interview results was based on records and researcher notes.

Table 3.3 The main questions for socio-environmental activists[a]

Tan Malaka's values	Interview questions
(1) Rejection to direct adoption of western (foreign) styles in the government's management system (NPM Styles)	1. What is your view about local governments' accountability policies, measurement-standards, activities, and official reports, specifically related to socio-environmental accountability issues in current days? 2. Have those accountability instruments been in accordance with the context faced by Indonesian people, specifically, related to the socio-environmental issues that you support now? 3. Are those accountability instruments bringing certain ideologies that tend to exacerbate the socio-environmental issues that you advocate? 4. Do you have any suggestions for development of local government's accountability activities and report, specially related to the socio-environmental issues that you advocate now?
(2) The importance of representation of every class and group in the community to be given a place in the [process of] building a government Accountability system	1. What is your view about the representation of people's voice in local governments' accountability policies, activities, and official reports, specifically related to the socio-environmental accountability issues in nowadays? 2. Are all groups of society, specifically those currently facing socio-environmental problems, involved in the making of socio-environmental policies by the government? 3. How far is your campaign value involved in government's socio-environmental policies and accountability? 4. Do you have any suggestions for the representation of the society's values in the government's accountability activities, measurement-standards and reports, particularly, related to the socio-environmental issues that you advocate now?

(continued)

Table 3.3 (continued)

Tan Malaka's values	Interview questions
(3) Focus on how social structures and agent (individual/society) can influence each other	1. How far is the society's involvement in the local governments' accountability policies making, particularly, related to the socio-environmental issues that you advocate now? 2. Is there any approach by the government so its socio-environmental policy meets the society interest? As another form of government's accountability, specially, related to socio-environmental issues on your concern now 3. Do you have any suggestions so that the government's interest can accommodate people's interest in socio-environmental accountability issues?

[a]As this research used a semi-structured interview style, the main questions above were developed for further questions naturally during the interview

2. *The Government's Officials*

Moreover, in order to make the informant's data to be more reliable, the triangulation approach was used (Denzin & Lincoln, 2005). The semi-structured interview also applied to government officials who [write/submit] and review the government accountability reports. These open interviews to government's officials will focus on how they prepare their accountability activities and reports (in local governments[8]), and how they review or oversee the accountability practices in the government agencies (the internal auditory board of local governments[9]), especially, within the particular areas in this research setting.

The interview step was intended to be done with the governments' officials, started with those from Makassar City and South Sulawesi Province who compile the government accountability reports. However, there was a significant obstacle on the interview arrangement with informants from the internal auditory board (inspectorate) from both local governments who review and oversee the accountability practices in the government agencies. All of them politely refused my request. One official of the auditory board said:

> For city government's accountability report, we simply do an official review based on standard designated byKEMENPAN-RB.[10] So, no special information you can get in this office, I suggest you to go to KEMENPAN-RB to obtain a deeper understanding on government's accountability system.

[8]Provinces, Cities, Districts government.

[9]BPKP (Board of Development and Financial Control) and/or Inspectorate of local governments.

[10]*Kementerian Pendayagunaan Aparatur Negara dan Reformasi Birokrasi Republik Indonesia* (Ministry of Administrative and Bureaucratic Reform of the Republic of Indonesia).

Table 3.4 Summary of informants from government side

Codes of informants	Social-environmental cases	Cities/Districts	Category of informants	Number of informants (FGD)	Time
GOS1	Coastal Reclamation	Makassar, South Sulawesi	South Sulawesi Provincial Government	1	2/1/19 (±60 min)
GOM1			Makassar city Government	1	5/1/19 (±60 min)
GOS1	Environmental issues in the mining area	Galesong, South Sulawesi	South Sulawesi Provincial Government	1	2/1/19 (±60 min)
MA1	Ministry of Administrative and Bureaucratic Reform as the **standard setter** and **supervisor** of Government's Accountability System			1	1/3/19 (±60 min)

Taken that situation into account and in order to have more complete data from the government side, an additional interview in the form of Focus Group Discussion (FGD) with the standard setter's officers of the system of performance accountability in government agencies (SAKIP)[11] in KEMENPAN-RB[12] was organized. However, since several officers were highly occupied, FGD was not running well. Then I changed to interview a senior expert staff in KEMENPAN-RB. The interview focus was to elaborate how the ministry develops, implements and oversees the rules of accountability activities in government agencies, specifically, the attention of SAKIP on socio-environmental issues. The following is the list of my main informants from the government side (Table 3.4).

Furthermore, in line with the intention to harmonize among the value of research methodology and method, the following questions were asked for the government officials (Table 3.5).

3.4.1.2 Documentation

This research focuses on criticize and develop local government's accountability framework. Therefore, this research used supporting documents to discuss and synthesize the research objective. From the government side, this research used the government accountability documents such as the fundamental laws, Financial Statements and the Performance Accountability Reports of Government Agencies (LAKIP), specifically, the accountability documents from regions that become

[11] *Sistem Akuntabilitas Kinerja Instansi Pemerintah (SAKIP).*

[12] *Kementerian Pendayagunaan Aparatur Negara dan Reformasi Birokrasi Republik Indonesia* (Ministry of Administrative and Bureaucratic Reform of the Republic of Indonesia).

research settings of this study. This research also used the central discourse related to the United Nation's Sustainable Development Goals (SDGs). In that regards, SDGs documents are also used as supporting documents. In addition to that, other frameworks for accountability reports related to sustainability issues become additional documents to enrich the research data. Finally, yet importantly, previous studies related to governments accountability and comparative data from the mass media were also used.

Table 3.5 The main questions for the government officials[a]

Tan Malaka's values	Interview questions
(1) Rejection to the direct adoption of western (foreign) styles in the government's management system (NPM Styles)	1. What is your view about local governments' accountability policies, activities, measurement-standards, and official reports, specifically related to the current socio-environmental accountability issues in your region? 2. In your opinion, have those accountability instruments been in accordance with the context faced by Indonesian people, specifically, related to socio-environmental issues in your region now? 3. Are those accountability instruments bringing certain ideologies or business styles to the government accountability field? 4. Do you have any suggestions for future development of local government's accountability activities and report? Primarily, related to the socio-environmental issues in your region
(2) The importance of representation of every class and group in the community to be given a place in the [process of] building a government Accountability system	1. What is your view about the representation of people's voice in local governments' accountability policies, activities, and official reports, specifically related to the current socio-environmental accountability issues? 2. Are all groups of society, specifically those currently facing socio-environmental problems, involved in the making of socio-environmental policies of the government? *(for government officials only)* 3. Do you have some inputs about representation of society's values in government's accountability activities, measurement-standards, and reports? Mainly, related to the socio-environmental issues in your region nowadays

(continued)

Table 3.5 (continued)

Tan Malaka's values	Interview questions
(3) Focus on how the social structures and agent (individual/society) can influence each other	1. How far is the society involvement in the local governments' accountability policies making? Particularly, related to the socio-environmental issues in your region *(for government officials only)* 2. Is there any approach by the government bodies so their socio-environmental policy meets the interest of the society? As another form of government's accountability. Especially, related to the socio-environmental issues that become public focus in your region? 3. Do you have some inputs so the government's interest can accommodate the people's interest in socio-environmental accountability issues?

[a]As this research used a semi-structured interview style, the main questions were developed for further questions naturally during the interview

3.4.1.3 Observation

Lastly, in order to gain essential information from the people and environment directly affected by the socio-environmental issues, the focused observation and FGD were used as an approach to collect data. This observation focuses to distinguish certain aspects: how the involvement of local people in local governments' socio-environmental accountability policymaking, what their real situations are like to face the socio-environmental issues, and specifically related to the position of the people [voices and conditions] in the government's accountability activities, as in line with the Tan Malaka's values that used in this research. The Galesong district and its people in a forum called *Forum Masyarakat Pesisir-Nelayan Galesong Raya (FORMASI-NEGARA)*[13] become the main local guidance in observation and FGD activities. Galesong district is one of the coastal areas experiencing social problems and environmental damage due to a reclamation project in Makassar. Galesong's coastal area is used as quarry of sand material for Makassar reclamation project.

[13]A forum of coastal communities and fishermen in Galesong district, Takalar Regency, South Sulawesi.

3.4.2 Data Analysis

After collecting the summary of information from the data collection steps, at this stage, the summary of the research data then categorized by using NVIVO-12 software, analyzed, and synthesized with Tan Malaka's Critical-Emancipatory Methodology as developed for this research. For more details, steps taken in data analysis that comes from the direct translation of Tan Malaka's methodology are described as follows (Fig. 3.3).

These logical research steps are built on this research's methodological coherence. Therefore, starting from the collection data, I put a reference of the Tan Malaka's critical-emancipatory methodology values[14] and this study focused on the directed-way, a way to criticize and to develop the local government's accountability framework to be more emancipative and integrated.

In addition to that, the synthesis action of this research is based on the data collections. Specifically, it is based on main aspirations from Socio-environmental activist's critical views, the government's practices, and the main points of SDGs. The data collection summary was classified by Tan Malaka values and two focuses of Tan Malaka's critical-emancipatory methodology, namely structural and non-structural synthesis. Structural synthesis focus on data collection, summarize, and discussion related to the structural development of accountability system. In contrast, non-structural synthesis, or in this research is called as gathering society information, emphasis on gathering and arguing the data related to the informants' information and additional documents that were used as main ideas to develop a new integrated accountability framework in local governments. The coherence between these two synthesis processes becomes inseparable at this stage.

In the end, the points of synthesis results are categorized into the new form of accountability framework that reflects the value-dimensions of economic, social, and environmental elements. Hopefully, it could create a more emancipative local government accountability framework for society's interest and be more comprehensively accountable to the Indonesian people. It could also be an alternative reference for countries that also find it challenging to manage their socio-environmental problems.

[14]Look again at the list of questions to informants in Sect. 3.4.1.

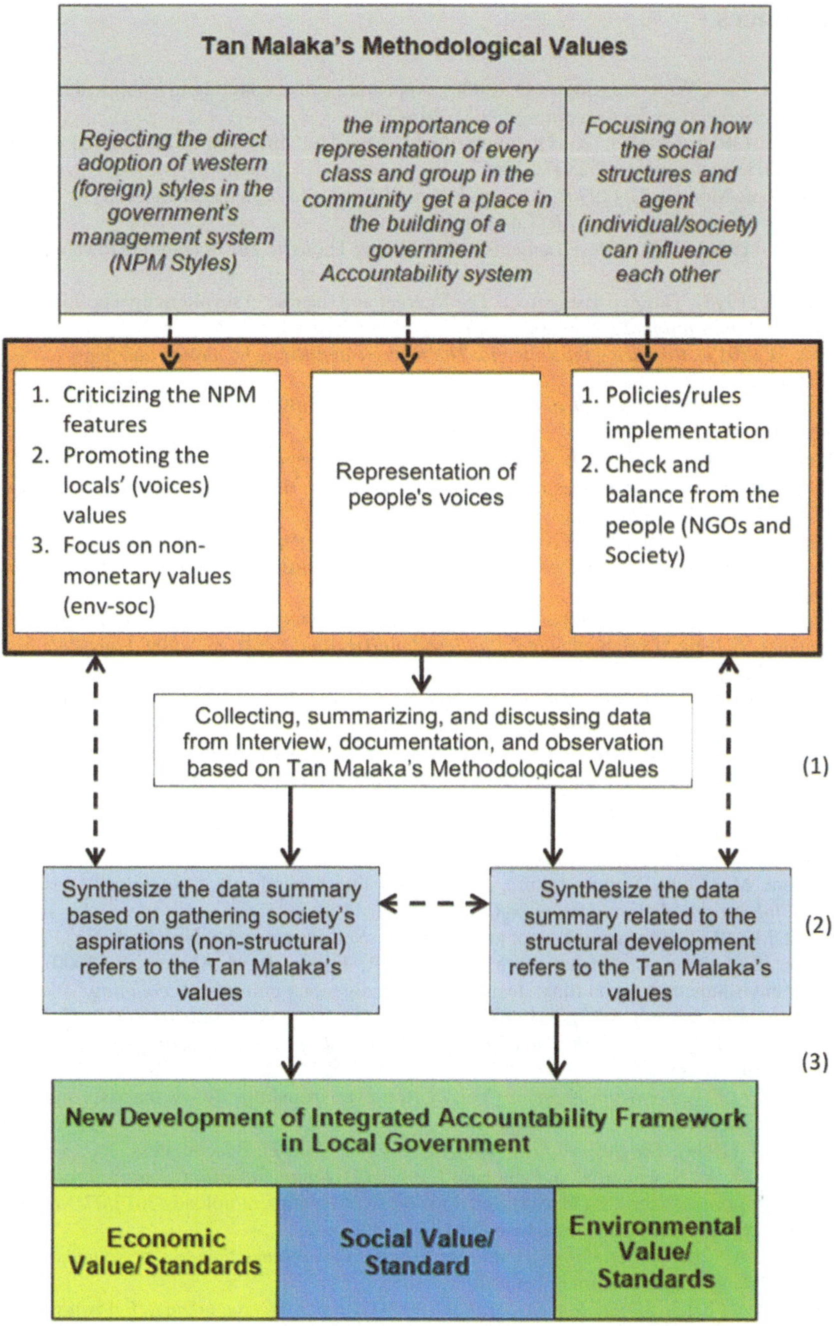

Fig. 3.3 Steps in data analysis

References

Alawattage, C., & Wickramasinghe, D. (2008). Appearance of accounting in a political hegemony. *Critical Perspectives on Accounting, 19*(3), 293–339.

Bhabha, H. (1984, October 28). Of mimicry and man: The ambivalence of colonial discourse, 125–133. https://doi.org/10.2307/778467.

Burrell, G., & Morgan, G. (1979). *Sociological paradigms and organisational analysis: Elements of the sociology of corporate life*. London: Heinemann.

Chua, W. F. (1986). Radical Developments in Accounting Thought. *The Accounting Review, LXI*(4), 601–632.

Cleaver, H. (1998). The Zapatista effect: The Internet and the rise of an alternative political fabric. *Journal of International Affairs*, 621–640.

Cleaver, H. (2017). *Rupturing the dialectic: The struggle against work, money, and financialization*. AK Press.

Creswell, J. W. (2013). *Research design: Qualitative, quantitative, and mixed methods approaches*. Sage.

da Fonseca, A. C. P. D. (2010). Management control systems: Universal practices or national practices? *Critical Perspectives on International Business, 6*(2/3), 190–207. https://doi.org/10.1108/17422041011049996.

Denzin, N. K., & Lincoln, Y. S. (2005). *The Sage handbook of qualitative research*. Sage.

Diamond, J. (2013). *The world until yesterday: What can we learn from traditional societies?* Penguin.

Diefenbach, T. (2009). New public management in public sector organizations: The dark sides of managerialistic "Enlightenment." *Public Administration, 87*(4), 892–909. https://doi.org/10.1111/j.1467-9299.2009.01766.x.

Djamhuri, A. (2009). *A case study of governmental accounting and budgeting Reform at local authority in Indonesia: An institutionalist perspective*. USM.

Duarte, C. P., da Fonseca, A., & da Fonseca, A. C. P. D. (2010). Management control systems: Universal practices or national practices? *Critical Perspectives on International Business, 6*(2/3), 190–207. https://doi.org/10.1108/17422041011049996.

Escobar, A. (1998). Power and visibility: Development and the invention and management of the Third World. *Cultural Anthropology, 21*(4), 428–443.

Finnemore, M. (1996). Norms, culture, and world politics: Insights from sociology's institutionalism. *International Organization, 50*(2), 325–347.

Foucault, M. (1972). The archeology of knowledge (A. Sheridan, Trans.). New York, NY: Pantheon.

Gallhofer, S., Gibson, K., Haslam, J., McNicholas, P., Takiari, B., & Authors, F. (2000). Developing environmental accounting: Insights from indigenous cultures. *Accounting, Auditing & Accountability Journal, 13*(3), 381–409. https://doi.org/10.1108/09600030410545445.

Giddens, A. (1984). *The constitution of society: Outline of the theory of structuration*. University of California Press.

Gramsci, A. (1994). *Letters from prison* (Vol. 2). Retrieved from https://www.marxists.org/archive/gramsci/index.htm.

Habermas, J. (1984). *The theory of communicative action* (Vol. 2). Beacon press.

Kaltenbrunner, A., Newman, S., & Painceira, J. P. (2012). *Financialisation of natural resources: A marxist approach*. Retrieved from https://iippe.org/wp/wp-content/uploads/2013/07/Annina-Kaltenbrunner_Susan-Newman_and-Juan-Pablo-Painceira.pdf.

Kamayanti, A. (2016). *Metodologi Penelitian Kualitatif Akuntansi: Pengantar Religiositas Keilmuan*. Malang: Yayasan Rumah Peneleh.

Kamayanti, A. (2017). *Metodologi Konstruktif Riset Akuntansi: Membumikan Religiositas*. Malang: Yayasan Rumah Peneleh.

KPA. (2017). KPA Launching Catatan Akhir Tahun 2016. Retrieved April 17, 2017, from https://www.kpa.or.id/news/blog/kpa-launching-catatan-akhir-tahun-2016/.

Krishnaswamy, R. (2002). The criticism of culture and the culture of criticism: At the intersection of postcolonialism and globalization theory. *Diacritics, 32*(2), 106–126.
Kristiansen, S., Dwiyanto, A., Pramusinto, A., & Putranto, E. A. (2009). Public sector reforms and financial transparency: Experiences from Indonesian districts. *Contemporary Southeast Asia: A Journal of International and Strategic Affairs, 31*(1), 64–87. https://doi.org/10.1355/CS31-1C.
Latif, Y. (2011). *Negara paripurna: Historisitas, rasionalitas, dan aktualitas Pancasila.* Gramedia Pustaka Utama.
Laughlin, R. C. (1987). Accounting systems in organisational contexts: A case for critical theory. *Accounting, Organizations and Society, 12*(5), 479–502.
LBH-Jakarta. (2015). *Atas Nama Pembangunan: Laporan Penggusuran Paksa di Jakarta Tahun 2015* (Vol. 2017). Retrieved from https://www.bantuanhukum.or.id/web/atas-nama-pembangunan-laporan-penggusuran-paksa-di-jakarta-tahun-201.
LBH-Makassar. (2016). Siaran Pers Advokasi Masyarakat. Retrieved from https://lbhmakassar.org/siaran-pers/.
Lehman, G. (2001). Reclaiming the public sphere: Problems and prospects for corporate social and environmental accounting. *Critical Perspectives on Accounting, 12*(6), 713–733. https://doi.org/10.1006/cpac.2001.0476.
Llorente, R. (2006). Analytical marxism and the division of labor. *Science & Society, 70*(2), 232–251.
Macintosh, N. B., Hopper, T., & Figure, S. (2005). *Accounting, the social and the political: Classics, contemporary and beyond*. https://doi.org/10.15713/ins.mmj.3.
Malaka, T. (1921). *Parlemen atau Soviet* (2014, Ed.). Jakarta: Sega Arsy.
Mayer, M. (2008). *Anarchism for beginners*. Red Wheel/Weiser.
Mulawarman, A. D. (2010). Integrasi Paradigma Akuntansi: Refleksi atas Pendekatan Sosiologi dalam Ilmu Akuntansi. *Jurnal Akuntansi Multiparadigma, 1*(1), 155–171.
Mulawarman, A. D. (2016). *2024 Hijrah Untuk Negeri: Kehancuran atau Kebangkitan Indonesia dalam Ayunan Peradaban*. Jakarta: Yayasan Rumah Peneleh.
Nezar, P., & Andi, A. (2009). *Antonio Gramsci: Negara dan Hegemoni*. Yogyakarta: Pusataka Pelajar.
Poeze, H. A. (2008). *Tan Malaka, Gerakan Kiri, dan Revolusi Indonesia 1*. Jakarta: Yayasan Obor Indonesia.
Powell, J. (1998). *Postmodernism for beginners*. New York, USA: Writers and Readers Publishing Inc.
Rajiani, I. (2011). A cultural ecology of new public management in Indonesia. *Journal of Administrative Science, 8*(1), 17–31.
Ritzer, G., & Jurgenson, N. (2010). Production, consumption, prosumption: The nature of capitalism in the age of the digital "prosumer." *Journal of Consumer Culture, 10*(1), 13–36. https://doi.org/10.1177/1469540509354673.
Ritzer, G., & Stepnisky, J. (2017). *Modern sociological theory*. Sage.
Ryer, R. A. B. (1999). A Marxist critique of the Fasb' S conceptual framework. *Critical Perspectives on Accounting*, 551–589.
Said, E. W. (2012). *Culture and imperialism*. Vintage.
Sarantakos, S. (2012). *Social research*. Macmillan International Higher Education.
Saravanamuthu, K., & Lehman, C. (2013). Enhancing stakeholder interaction through environmental risk accounts. *Critical Perspectives on Accounting, 24*(6), 410–437. https://doi.org/10.1016/j.cpa.2013.02.002.
Silverman, D. (1985). *Qualitative methodology and sociology: describing the social world*.
Spence, C. (2009). Social accounting's emancipatory potential: A Gramscian critique. *Critical Perspectives on Accounting, 20*(2), 205–227. https://doi.org/10.1016/j.cpa.2007.06.003.
Springer, S. (2012). Neoliberalism as discourse: Between Foucauldian political economy and Marxian poststructuralism. *Critical Discourse Studies, 9*(2), 133–147. https://doi.org/10.1080/17405904.2012.656375.
Stalin, J. V. (1978). On the final victory of socialism in the U.S.S.R. In *Works, Vol. 14* (Internet A). Retrieved from https://www.marxists.org/reference/archive/stalin/works/1938/01/18.htm.

Stiglitz, J. E., Sen, A., & Fitossi, J.-P. (2009). *Commission on the measurement of economic performance and social progress report-2009*. Paris: The Commission on the Measurement of Economic Performance and Social Progress-France.

Tinker, A., & Gray, R. (2003). Beyond a critique of pure reason: From policy to politics and praxis in environmental and social research. *Accounting, Auditing, and Accountability Journal, 16*(5), 727–761.

Toft, K. H. (2015). Liberal CSR and new marxist criticism. In *Corporate social responsibility and governance* (pp. 303–316). Springer.

Trotsky, L. (2010). *The permanent revolution & results and prospects*. Red Letter Press.

UNDP. (2018). Sustainable development goals. Retrieved May 15, 2018, from United Nations Development Programme website: https://www.id.undp.org/content/indonesia/en/home/sustainable-development-goals.html.

WALHI. (2015). *Tinjauan Lingkungan Hidup 2015 WALHI*. Retrieved from https://walhi.or.id/wp-content/uploads/2016/12/OutLook2015_W.pdf.

WALHI. (2016). *Tinjauan Lingkungan Hidup WALHI 2016: Keharusan Pembenahan Struktural untuk Perbaikan Tata Kelola*. Retrieved from Wahana Lingkungan Hidup Indonesia website: https://www.walhi.or.id/2016/12/22/outlook-2016-tujuan-lingkungan-hidup/.

Watchdoc. (2018). Kumpulan film dokumenter kerusakan lingkungan di Indonesia. Retrieved February 14, 2018, from Youtube website: https://www.youtube.com/user/watchd0c/featured.

Chapter 4
Description of Study Results

The subaltern "cannot speak," instead,
because their speech falls short of fully authorized, political speech.
Too much gets in the way of their message's being heard, socially and politically.
—Gayatri Chakravorty Spivak (Indian Critical Philosopher)
Changes and progress very rarely are favors from above.
They come out of fights from below.
—Noam Chomsky (American Critical Philosopher)

The previous chapters discussed about the research methodology based on Tan Malaka's critical views, and how to apply it in this research scheme to develop an integrated [emancipative] framework of local government accountability that integrates economic-socio-environmental issues based on aspirations of Indonesian socio-environmental activists, society, as well as supporting the achievement of sustainable development goals. Now, this chapter is to discuss about study results from (1) aspirations of Indonesian socio-environmental activists who are actively struggling to advocate the government's accountability on socio-environmental issues, (2) the affected people by a socio-environmental case in evaluating the local government's accountability related to that issue, and, (3) government officials who are responsible to prepare the local government accountability report.

In order to gain a deeper understanding, I also triangulate the data collected from interview and observation with government's regulations, published reports, and media news. I believe that by using the triangulation technique, helps me to more light up the views offered by my informants (Belal, Cooper, & Khan, 2015; Creswell, 2007). Moreover, for this research, I believe, the real emancipative action are to discuss and analyze society aspirations related to the voices of Indonesian socio-environmental activists who are actively fighting to advocate the government's accountability on socio-environmental issues in Indonesia, and how the affected people evaluate the local government's accountability related to their socio-environmental issue.

H. M. Shahib, *Towards the Local Government's Integrated Accountability Framework*,
https://doi.org/10.1007/978-981-33-6617-6_4

4.1 Government Accountability: Amid the Implementation and Socio-Environmental Issues in South Sulawesi

Here I indicate examples for aspirations of people who are not being accommodated in accountability practices, specifically in socio-environmental issues that have a wide impact to the society. One of the examples are the coastal reclamation case in Makassar city of Indonesia's South Sulawesi province which has also affected Galesong district and fishermen in the area.

Based on this research's observation and interview with local people in FORMASI NEGARA group, Galesong coast become a quarry for sand material to hoard the reclamation land in coastal area of Makassar city. This situation damages ecology in Galesong's coastline. For example, abrasion damages the coastal and fishermen's residential area, deepening of fishing area, and the fish's ecosystem damage. As the effect of these environmental problems, Galesong's fishermen livelihood and activities are threatened. Currently, they cannot go fishing nearby Galesong area. They must go far away as there is no fish anymore around the district's coastline. These conditions make these fishermen suffer double losses. They suffer economic losses, due to increasing operational fishing costs, and environmental damage in the area where they live (Pictures 4.1 and 4.2).

Referring toIndonesian Government's development plan and performance accountability guidelines—The Government Regulation of the Republic of Indonesia Number 8/2008 About Steps, Procedures for Preparation, Control and Evaluation of

Picture 4.1 Abrasion effect on Galesong's public cemetery located near to the coastline

Picture 4.2 Abrasion effect on a house located near to the coastline in Galesong (Photos were taken from researcher observation in Galesong District. To know more deeply the sand quarry's effect on the Galesong environment coastline because of the land reclamation in the coastal area of Makassar city, please read the following in-depth reporting (Environmental Justice Atlas, 2019; Mongabay, 2017b, 2020)

Regional Development Plan Implementation,[1] and The President Regulation Number 29/2014 as the legal basis for the system of performance accountability in government agencies *(SAKIP)*—these two regulations are the basis that bind the development plan, policy implementation, and accountability of government performance top down. One thing that makes these two regulations intersect is the strategic plan determination step.

Strategic plan determination step is an important step for all local governments in Indonesia. This is a vital step because it is the map for government's works (according to the Government Regulation of the Republic of Indonesia Number 8/2008) and it also the reference to make the governments' accountability performance indicators (according to The President Regulation Number 29/2014). The interesting point is, in making the government's strategic plan that integrated with the RPJMD[2] (a short-term regional development planning document). There is a regulated step that known as *Musrenbang*.[3] The inter-stakeholder[4] forum in the context of preparing regional

[1]In Bahasa Indonesia: *Peraturan Pemerintah Republik Indonesia Nomor 8 Tahun 2008 tentang Tahapan, Tata Cara Penyusunan, Pengendalian dan Evaluasi Pelaksanaan Rencana Pembangunan Daerah.*

[2]In Bahasa Indonesia RPJMD: *Rencana Pembangunan Jangka Menengah Daerah,* a regional development planning document for a period of 5 (five) years.

[3]In Bahasa Indonesia Musrenbang: *Musyawarah Perencanaan Pembangunan.*

[4]Stakeholders are those who directly or indirectly gained benefit or impact from the planning and implementation of regional development. These explanations refer to the Government Regulation of the Republic of Indonesia Number 8/2008.

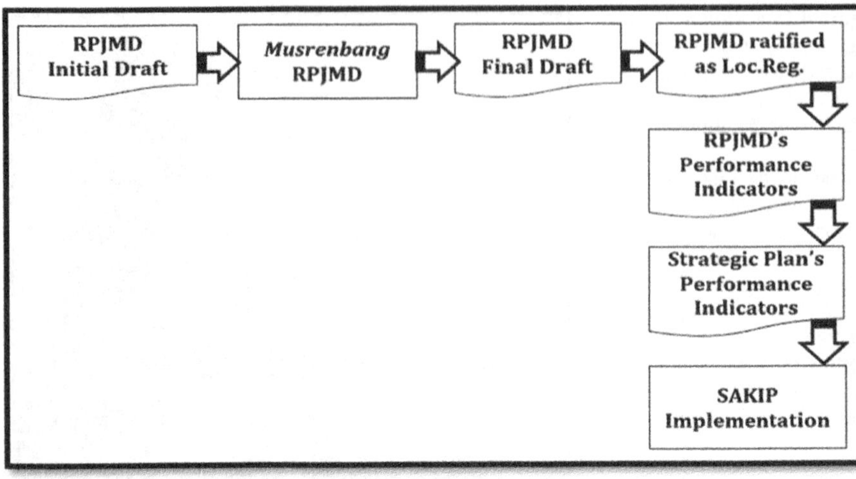

Fig. 4.1 The Intersection between RPJMD and SAKIP based on the government regulation of the republic of Indonesia Number 8 of 2008 and The President Regulation Number 29/2014

development plans before the RPJMD ratified as the local government's regulation and become the standards for governments' accountability performance indicators (see Fig. 4.1).

If the intersection between RPJMD and SAKIP is associated with Galesong people, who affected with the sand mining activities to hoard Makassar's reclamation land, Here, It can be seen a clear example of a gap between what should the local government do in regards to regulations, and what is happening to the Galesong people. Hence, this situation is reflected in government's policies related to Makassar's coastal reclamation project up to accountability activities and report which are done by local governments. For example, as our informant explained in the following words:

> There was never such a thing [community involvement]. However, [maybe] the village head was always involved, and from the village head the information was interrupted, because the village head knew that if it was delivered to the people, the people would definitely reject the program and it will not operate anymore. It is because that people automatically know if there is a [sand] mining, the fish will run away [from Galesong's sea] and nothing else can be caught. [Galesong] people were never involved to talk about it (FN1).

Based on the local informant's explanation above, I see there was nothing like *Musrenbang* that involve local people in Galesong to discuss related to reclamation project and its consequence in the form of sand mining in Galesong district before the project runs. Although, *Musrenbang* must be done as per local government's development planning program, and Galesong people are the government's stakeholder, who are directly affected by the planning and implementation of regional development, refers to the Government Regulation of the Republic of Indonesia Number 8/2008.

Furthermore, as I see this condition in correlation to government's accountability practices under SAKIP, I also did not find any explanation regarding consequences of current Makassar's reclamation project (CPI)[5]–which started in 2013 (Ruslin, 2017) to today (read: to 2019, when I wrote the results of this study)–in South Sulawesi provincial's accountability reports (LAKIP) in period of 2014–2018, despite the fact that south Sulawesi province government was the initiator and regulator of this big project. Thus, I argue that there are two possible explanations for this condition.

First, following the regulation logic, the void of information about the Makassar's reclamation project in South Sulawesi provincial's accountability reports (LAKIP) period 2014–2018 is the result of no-involvement of Galesong people to prepare the regional development planning program in the form of Musrenbang. Consequently, there is no government's performance indicators and explanation related to the Galesong's environmental damage in South Sulawesi provincial's accountability reports (LAKIP). Second, in defining RPJMD and government performance indicators, which become the standard of accountability system in government agencies, there is allegedly a strong institutional influence in official regulations and standards that guide the government. Thus, the people's problems as Galesong people have do not fall into the provincial government's mandatory things in their accountability reports.

These two possible explanations are more understandable as we refer to government regulations that become the basis in the preparation of RPJMD and government's performance indicators. For instance, in the regulation of the Minister of Home Affairs Number 54/2010 about implementation of the Government Regulation

[5]CPI or Center Point of Indonesia is the newest reclamation project for Makassar's coastline. It is initiated by the Government of South Sulawesi through Governor's Decree No. 644 year 2013 regarding the granting permit for coastal reclamation in Makassar-Losari's Coastline. CPI also cannot be separated from the central government program. The President of Indonesia in 2009 stated the Makassar City as a national strategic area with a planned project to build a state guesthouse (Wisma Negara) on the coast of Makassar. For CPI development, it is estimated that the area is acquired for the construction of the reclamation megaproject covering an area of 157 hectares. 57 hectares are allocated for the development of public space. More than 100 hectares currently are owned by Ciputra Group developers through Ciputra Surya Plc. in cooperation with Yasmin Bumi Asri Inc., to develop business areas, hotels, and real estate (Chandra, 2016; Ruslin, 2017). As the consequence of this megaproject, this project requires a lot of sand and soil, and one of the main quarries of sand is Galesong district area. For the photos of CPI project, the photos can be seen directly on:

(1) https://makassar.tribunnews.com/2019/01/11/foto-drone-reklamasi-kawasan-cpi-makassar-rampung.

(2) https://properti.kompas.com/read/2017/09/19/152857421/2019-ciputra-memulai-proyek-garuda-raksasa-di-lahan-reklamasi?page=all.

(3) https://ejatlas.org/conflict/reclamation-project-makassar-indonesia.

of the Republic of Indonesia Number 8/2008 about phase, procedures for preparation, control, and evaluation of the regional development plan implementation,[6] it shows that the government's performance indicator in RPJMD is based on the government priority programs that are aligned with the vision and mission of the elected Governor (the regulation of Minister of Home Affairs number 54/2010, p: 81–82). In this circumstance, there is an opportunity for the influence of political power or political lobbying in policy making to determine government's performance indicators in the RPJMD. It is because this indicator directly measures the successfulness of the governor's leadership period.

The situation above also was confirmed by the opinion of an informant from WALHi South Sulawesi (WALHI-SULSEL), who often received invitations to attend *Musrenbang* or consultative hearings to prepare draft regulations or regional policies as follows:

> For example, [public hearing] of local regulations, most public hearings are initiated by the government… However, unfortunately, when the government sent the invitations to us as CSO (Civil Society Organization), the drafts of regional regulation mostly were not attached in the official invitations. Even if they (government) attached [the drafts], some essential attachments were never included. For example, in the case of a regional regulation draft of spatial plan or a coastal zone zoning plan, [it is related to the reclamation case and sea sand mining in Galesong.] They (government) have never attached the coastal zoning map that will be regulated in the regulation [draft]… I mean, they made a draft regulation, and then we [CSO] were invited to criticize and give some input on the draft. Yet, they never gave the revised draft, how was the change in the draft that occurred after the public hearing, and the most fatal, it was before they (government) officially ratified [a regulation]. They (government) also never sent the [final] draft document to the public. It took a long time and process to be able to get the final draft to be ratified as a local regulation … [From this case, it can be seen] involving the community only to fulfill formal obligations, as though that the mechanisms and models of natural resource management systems had changed. However, it is not. People were involved, but their input was not accommodated, what I said as a pseudo-participation … as a result, none of the points from us (CSOs) were accommodated in the policy draft [of government]. [For the government,] the important thing was that there was WALHI's signature in the attendance list [as a formal proof of attendance] (WSS1).

Moreover, as accountability reporting is a compulsory-annual work based on the ratified RPJMD's performance standards as the local regulation (see again Fig. 4.1), it becomes an authoritative reference in implementation of the accountability system (SAKIP). There is also a performance agreement between regional heads and subordinates related to performance achievement based on the government's strategic plan that is integrated with the RPJMD. It directly makes all government agencies and employees focus on achieving the indicators they agreed too, as the government's performance indicators also relate to measuring their job performance. These conditions were confirmed by our informants from the government's agencies who have a responsibility in preparing the city and provincial government's accountability report, as below:

---

[6] In Bahasa Indonesia: *Peraturan Menteri Dalam Negeri Nomor 54 Tahun 2010 Tentang Pelaksanaan Peraturan Pemerintah Nomor 8 Tahun 2008 Tentang Tahapan, Tatacara Penyusunan, Pengendalian, Dan Evaluasi Pelaksanaan Rencana Pembangunan Daerah.*

> So we made the [accountability] report related to my job or my main duty, which was to create a report that had a prior performance agreement (GOS1).
>
> The accountability that we measure is the accountability we promised in the planning document. All programs, activities, targets in the planning document that have been set in the [RPJMD] local regulation, are the [government] achievement indicators (GOM1).

Interestingly, the informant also added that the indicators used as references in local governments' performance agreements and accountability reports, tend to use macro and monetary references. As stated by the informant:

> So the issues that are directly in contact with the people in the report are not very visible because [the provincial accountability report] is too macro (GOS1).

And as shown in the following accountability report of the South Sulawesi provincial government, the report refers to the regulation of the Minister of Home Affairs number 54/2010 about implementation of the Government Regulation of the Republic of Indonesia Number 8/2008 (about phases, procedures for preparation, control, and evaluation of the regional development plan implementation), and President Regulation Number 29/2014 about Accountability System of Government agencies' performance.

Based on the explanation above, an accountability system implementation is primarily related to the political influence because the indicator targets achievement will ultimately lead to an assessment of a regional head regime's success. The

Table 4.1 The example of RPJMD's indicators in accountability report (LAKIP-2017) of government of South Sulawesi Province

No.	RPJMD goals		Key performance indicators	Performance target (Realization)	%
1	To maintain a conducive situation to religious life	1	Number of social conflicts related to religious life	0 (0)	100
2	To bring a good economic growth into reality	2	Percentage of Economic Growth	7,3 (7,23)	97,7
		3	GRDP (Groos Regional Domestic Product) Per Capita	52,79 (48,21)	91,3
		4	Inflation	5,3 (4,17)	83,4
		5	Unemployment percentage	4,9 (5,61)	87,3
		6	Gini Ratio	0,4 (0,42)	93,2
3	To materialize Social Welfare	7	Percentage of Poverty Rate	8,74 (5,61)	155
4	To maintain the quality of Regional Environment	8	Regional Environmental Quality Index	63,25 (64,80)	102,4
5	To increase the Quality of Human Life	9	Human Development Index (HDI)	70,97 (70,97)	100

(continued)

Table 4.1 (continued)

No.	RPJMD goals		Key performance indicators	Performance target (Realization)	%
		10	Average Number of Schooling years	7,84 (7,84)	100
		11	Life expectancy number	70,1 (70,62)	100,7
6	To increase the Infrastructure support	12	Percentage of infrastructure index	71,11 (71,11)	100
7	To upgrade the competitiveness of regional Investment	13	Total investment realization value (Foreign and domestic investment)	Rp 9 trillion (11, 4 trillion)	127
8	To increase the Gender Equality in various aspects of community life	14	Gender Development Index (GDI)	70,16 (92,79)	132,2
		15	Gender Empowerment Index (GEI)	93,46 (70,02)	74,9
9	To reduce public disruption, and maintain the peace and convience in public sphere	*16*	Crime Rate (Number of crimes reported)	19.706 (7.598)	259,3
10	To increase the governance quality on regional finance management	*18*	Opinion of Government's Financial Statement	Unqualified Opinion *WTP* *(WTP)*	100

accountability system is also related to the strong institutional influence in the form of regulations that guide the government. Then it can be understood why social, environmental issues such as the Makassar beach reclamation and its impacts on Galesong community were not presented in the final accountability report of the South Sulawesi province government.

Furthermore, these situations also indicate that the NPM (New Public Management) influence has been so strong in the practice of government accountability in Indonesia. It is starting from the regulations that govern it to the practices that occur.

4.2 The Strong Influence of NPM in Local Government Accountability Practices in Indonesia

Other evidence of NPM influence in Indonesian government accountability practices can be seen in accountability reports from several regional governments that also face social-environmental issues and intense protests from various elements of society, particularly from social-environmental NGOs. Table 4.2 then summarized it.

Table 4.2 The Socio-Environmental issues disclosure in local government accountability reports (LAKIP)

	Socio-Environmental issues in public														
	Reclamation					Mining conflict					Land-living Eviction				
	Year (20…)														
Province/Region/City	14	15	16	17	18	14	15	16	17	18	14	15	16	17	18
South Sulawesi[1]	NA	NA	NA	NA	NA	NA	NA	NA	NA	NA					
Makassar[1]	NA	NA	PD	NA	NA										
Bali[2]	NA	NA	NA	NA	NA										
Central Java[3]						NA	PD	PD	NA	NA					
Rembang[3]						PD	PD	NA	NA	PD					
Jakarta[4]	NA	NA	NA	PD	NA						NA	NA	NA	NA	NA

(continued)

Table 4.2 (continued)

	Socio-Environmental issues in public						
	Reclamation	Mining conflict	Land-living Eviction				
	Year (20…)						
Yogyakarta[5]			PD	PD	PD	PD	PD

Notes

In the year 2014 to 2018, in (1) South Sulawesi Province and its Makassar City, two socio-environmental issues that seized the public attention and had an extensive socio-environmental impact were cases of coastline reclamation in Makassar city and Sand mining in Galesong. In (2) Bali province, the Benoa Bay reclamation project was an issue that seized the people's attention for refusing it as Balinese saw the project having extensive socio-environmental effect to Bali. (3) In Central Java Province and its Rembang Region, the refusal of a cement mining project from local communities in Kendeng, Rembang, seized public attention. The public saw that the project has an extensive effect on the local environment and people's livelihoods. (4) In the Special Province of Jakarta, the reclamation of Jakarta's north coast and forced evictions to the poor people's land-living were the issues that seized public attention as both issues had an extensive socio-environmental impact. (5) In Yogyakarta Province, the socio-environmental impact of people's land-living eviction for the New Yogyakarta International Airport (NYIA) development project was an issue that seized public attention

*The issues sampling is based on NGO's social-media activism through hashtags (#) trend on social media. For instance, #MakassarTolakReklamasi (Makassar Reject Reclamation), #BaliTolakReklamasi (Bali Reject Reclamation), #RembangMelawan, #SaveRembang (for the case of rejection of a cement mining operation in Rembang), #JakartaTolakReklamasi (Jakarta Reject Reclamation), #SaveBukitDuri and #StopPenggusuran (for Jakarta evictions), #StopNYIA #SaveKulonProgo (For eviction cases of the construction project of NYIA, Yogyakarta)

FD: Full Disclosure

PD: Partial Disclosure

NA: Not Available

**Data collection based on LAKIP accessed through the official website of each local governments

On Table 4.2, it can be seen that uniformity of accountability practice presents in the local governments with social-environmental issues and object of public criticism. They tend to conceal things that become the main point of public criticism. Although, when we refer to the logic and philosophy of accountability practices, things that need more explanation to the public become essential to be accounted for and explained in society to maintain public legitimacy (Deegan, 2002).

Furthermore, this further strengthens evidence of the strong influence of institutionalized NPMs in regulations, policies, and the practice of Indonesian local governments' accountability. To understand it further, article from Diefenbach (2009) can become a reference to explain characteristics of NPM in governance practices in public sector (government agencies). Diefenbach (2009) explains that since NPM began in the late 1970s, the adoption of NPM has gone global. Besides that, it is believed to be able to adapt to the political-social-locality in many countries in the world, from Europe, America, and Asia-Africa (Harun, An, & Kahar, 2013; Hope Sr, 2001; Lee & Haque, 2006; Vicente Pina & Torres, 2003). There are several advantages of NPM, which continued to be popularized by its supporters, so it tends to win the battle of public discourse in the world. Referring to Diefenbach (2009), there are at least five main points why NPM is highly popular as a leading discourse in the public sector. Interestingly, these five points are also sensed in the public sector field in Indonesia.

4.2.1 Assumptions About the Business Environment and Strategic Objectives

First, NPM's assumptions on the business environment and strategic objectives tend to be based on self-adjustment with the active movement of market mechanisms (Diefenbach, 2007, 2009; Karp, 2005). This assumption emphasizes the importance of public sector organizations to respond and to adjust onto market changes (Diefenbach, 2007; Karp, 2005). For example, in today's context, it can be seen in terms of investment offerings and services that the government provides to the public. These things have several similarities with services offered by private sector or even with other state or local governments', for example, education services, health, transportation, and investment opportunities in various leading sectors (Lorenz, 2012).

Thus, if the government preserves its strategic objectives and business environment to follow the market trends, there is a possibility that the public and investors will abandon public services and investment sectors that do not follow the market trends. Then, this situation will result on economic decline and [regional] government revenues decrease. It also affects the government's operational practices. These things then cause the government perspective [slowly] will proceed into market-oriented and based on efficient-effective measures, which is measured by monetary values, as the reference to achieve strategic goals (Diefenbach, 2009; Lorenz, 2012).

Referring to the first assumption of NPM that the government should adjust its business environment and strategic objectives with the market orientation, this situation also can be seen in the public sector field of local government in Indonesia. Especially, when we look at the practice of accountability of local government performance as summarized in Tables 4.1 and 4.2 above, local governments tend to put macro indicators in the accountability report only based on efficient-effective and monetary measures.

As commonly known, practice cannot be separated from the rules that govern it. For instance, the Government Regulation of the Republic of Indonesia Number 8/2008 about Steps, Preparation Procedures, Control and Evaluation of Implementation of Regional Development Plan as a basis for the preparation of the RPJMD which also works as a standard to evaluate local government performance, and finally to be reported in LAKIP. This regulation requires regional development planning to be carried out based on conditions and potentials of each region, following the dynamics of regional and national development. Next, the plan is formulated in a transparent, responsive, efficient, effective, accountable, participatory, measurable, fair, and sustainable manner (in articles 2 and 3, Chap. 2, Principles of Regional Development Planning). With additional explanation, "efficient" is achieving specific outputs with the lowest input or lowest input with maximum output, and "effective" is the ability to achieve targets with the owned resources by the most optimal way or process.

An effective-efficient perspective based on the value of money and market mechanisms to determine strategic objectives of local governments is more clearly seen in the regulation of the Minister of Home Affairs number 54/2010 about implementation of the Government Regulation of the Republic of Indonesia Number 8/2008 (about phase, preparation procedures, control, and evaluation of the regional development plan implementation) and "President Regulation Number 29/2014 about the Accountability System of Government agencies' performance". These two regulations regulate the link between assessing government work performance based on the achievement of government's program performance indicators and the use of the budget, up to an explanation of their accountability in LAKIP. Interestingly, the regulation of the Minister of Home Affairs number 54/2010 places one example of indicators to measure the investment climate, which is the number of demonstrations related to political, economic, and work strike issues. However, the use of these indicators can be adjusted to the [political] needs of local government. Thus, as can be seen in Tables 4.1 and 4.2 above, the number of demonstrations of citizens who were affected by socio-environmental problems is not present in the government's accountability report. Although, there were so many protests that the people did (aljazeera.com, 2019; Mongabay, 2017a; SUARA.COM, 2018; WALHI, 2018, 2019).

4.2.2 Assumptions About Organizational Structure and Processes

The second assumption of NPM regarding organizational structure and process stands on the view that an excellent public sector organization is a flexible organization and does not have a complicated hierarchy. In other words, it can be said that the organization structure tends to be decentralized, from planning to implementation of activities. Therefore, it will speed up the process of decision making and action (Diefenbach, 2009). This view then leads to the need for formalization of work standards and management strategies. Thus, the performance result can follow and focus on the primary duties of each field (Diefenbach, 2009; Torsteinsen, 2012).

Seeing this second assumption of NPM, it also feels like that it has been running in Indonesia, too. Several examples indicate this situation. For instance, it can be said that the NPM implementation model in Indonesia was a consequence of the 1997–1998 economic crisis and 1998 reform movement. It was due to the encouragement from various world economic bodies, such as the World Bank and IMF, to Indonesian government to transform the mindset of government bureaucracy in order to resolve the government's problem in an effective-efficient manner. This encouragement was in fact needed to accelerate the improvement of the country's economy after being hit by the 1997–1998 economic crisis (Harun et al., 2013).

Thus, it can be witnessed, slowly the style of government governance has shifted from previously highly centralized (during the new order period), now has become more decentralized. The decentralization regime was also supported by the emergence of laws and government regulations on regional autonomy. These include, "Government Regulation of the Republic of Indonesia Number 8/2008 about Steps, Procedures for Preparation, Control and Evaluation of Implementation of Regional Development Plan" and "President Regulation Number 29/2014 about the Accountability System of Government agencies' performance". The two are examples of regulations that were born with the spirit of regional autonomy and decentralized government practices in Indonesia.

According to Diefenbach (2009), it must be acknowledged that the decentralization of government worked to a lower level brings some positive impacts on society. Local government decision-making would bring instant effects to the people affected on an issue because no need for prior approval from the central government. Also, problems related to local government agencies can be directly resolved internally within the agency.

However, several things become anomalous towards the implementation of this decentralized NPM pattern in the public sector in various parts of the world, including Indonesia. An example of the NPM anomaly, which often occurs in the implementation of decentralization schemes in the public sector organizations structure and processes, is the presence of a "new centralized model." It means that the government tends to be more decentralized in terms of technical and operational matters. However, in strategic matters, for example, strategic goals setting, planning, policies, performance targets, financial budgets, and information systems, are still tend to be very centralized following government rules or standards (Courpasson, 2000; Diefenbach, 2009).

Whilst NPM supporters might argue that these standards can be developed according to local needs, however, one thing becomes a new fact that decentralization in the form of regional autonomy in Indonesia presents a new centralized power in the regional area. It is centered on political actors who hold power in the local area (Aspinall, 2013). The condition of new centralized power is more supported by the rules, which provide space to adjust the local government's performance standard. This situation then opens opportunity for political lobbying among local leaders. It is because the performance indicator standard will be a standard to evaluate the achievement of one government regime. Thus, it can be understood, if local governments in Indonesia tend hinder the aspiration of socially vulnerable groups, such as the poor, fishermen, farmers, who become victims in several government's development projects, as reflected in their accountability reports (see again Tables 4.1 and 4.2).

The further anomaly in the application of the decentralization model as a consequence of NPM is that the bureaucracy and government's apparatus focus more to achieve their targets instead of getting into the substance of the bureaucracy, as a public servant. The government's employees are more preoccupied with matters related to supervisions, such as performance reports preparation, undergo audits and accreditation from government and external supervisory bodies, which have specific cycles (quarterly, semester, and annual) (Diefenbach, 2005, 2009; Kirkpatrick, Ackroyd, & Walker, 2005:115). This situation also occurs in local governments in Indonesia, as confirmed by this research informant:

> We are not profit-oriented, but [our] financial accountability and performance are monitored by all internal and external supervisory agencies. For example, if it is not accountable, we can be punished. We already have our accountability supervisors. Every quarter there is an inspectorate coming in [for checking]. Thus, all supervisors are always there, so when we are not accountable, those who execute [to punish us] (GOM1).

> …Now, in the SKP (local government work unit), what we are working on, it must be under the duties and promises agreed upon [with the leaders] … So, the employees work according to the [performance] agreement (GOS1).

By referring to the anomaly of the changes in organizational structure and processes brought by NPM, there are indeed positive changes brought by NPM, too. Specifically, it relates to decentralization and simplification of local government organizations hierarchy (Courpasson, 2000; Diefenbach, 2009). However, there is also a room for the centralization of new powers in the local context. When we view this from political standpoint, there are opportunities for local leaders' (politicians) interests and political compromise to influence government accountability practices. It is because the accountability's performance indicators will be the basis to evaluate the development achievement of a ruling regime. Also, the report-oriented work and supervisory patterns can unconsciously build a mechanical, work-oriented organizational culture based on standards (Kirkpatrick et al., 2005:115), without considering the other contexts and issues that are difficult to be measured by monetary value (Diefenbach, 2004; Michael, 2005). For example, in the case of environmental conflict and natural resources, these cases are essential to account as a consequence of the democratic governance logic that grows in countries with natural resources wealth.

4.2.3 Assumptions About Performance Management and Measurement Systems

Furthermore, (Diefenbach, 2009) explains that the third assumption of NPM related to performance management and measurement systems, stand on the view of best-practice industry standards applied to the public sector. Especially those are related to management information systems, audits, control systems, quality assessments, performance evaluations, employee assessments, benchmarking, and feedback reporting mechanisms. It is hoped that the implementation of these things can improve efficiency, quality, productivity, and motivation in public sector.

The theoretical expectations of NPM performance management assumptions are indeed present in the practice of public sector organizations throughout the world. The world government that implements NPM in its governance practices has reaped an increasing effectiveness and efficiency of its performance, productivity, and quality of services provided to the public (Harun et al., 2013; Lee & Haque, 2006; Torres, 2004; Torsteinsen, 2012). Moreover, all of them are very based on facts and rational evaluations, which also contribute to increase of government accountability reliability (Courpasson, 2000; Diefenbach, 2009; Harun et al., 2013; Torsteinsen, 2012).

Unfortunately, due to the very rational-based assessment of work, NPM approach related to performance management and measurement systems is still very much based on financial measures and technical aspects merely on things that are visible (tangible). This condition then [unconsciously] influences other essential matters which tend to be disfavored in management performance measurement. For example, evaluations of government intangible assets such as culture, social capital, quality of life, equity, participation, social impact, community practices, all of them tend not to be given the right place in the practice of assessing the NPM performance (Diefenbach, 2004, 2009; Kirkpatrick et al., 2005:67; Michael, 2005).

As conveyed by several scholars in the previous paragraph, which is also in line with this research finding in Indonesian local government, the effect of NPM style adoption makes local governments and their apparatus tend to focus on technical matters that can be quantified and measured financially. Local governments tend to be out of focus on things that are difficult to measure, such as social-environmental issues occured in their area. These conditions are confirmed by their report (see Tables 4.2

and 4.3), and results of the following interview to the government's employees, who are responsible to the government's accountability report:

> All these responsibilities must be measurable; how to measure them is through established indicators. The target determines the indicators, and the target is adjusted to the region's capability, for example, [adjusted to the capability of] HR (Human Resources) and budget. [Also], Community satisfaction, for example, it must be quantitative first…The qualitative matters must be quantitative first. For instance, the number of complaints in the [LAKIP] report. Yet, the theme of the complaints is not [in the report], because it is not an indicator… Anyway, anything qualitative is quantified first. That is the function of the indicator. The better indicators make us easy to measure performance. So, if the indicator is not clear, then it is difficult to measure it (GOM1).

> The performance agreement is the basis for us to make the report every year. However, the target tends to be adjusted annually. For example, this year is 90%, next year maybe 95%, so the change is in the target. In my experience in making a report [LAKIP], there is no space [for the latest socio-environmental issues] in the report. However, in my opinion, indeed, it is essential that it should be in the report. It is just because we are limited by the system [following the writing guidelines] … and we also got data [for LAKIP compilation] that has been processed [by related government's agencies] … Those [related agencies] think the essential things are achieving their targets. Even if not achieved, but at least close to [achieve the targets] … (GOS1).

From the explanation above, it can be concluded that NPM assumptions about performance management and measurement systems have penetrated onto the governance of Indonesian local government. Furthermore, the evidence can be seen from positive and negative effects. Positive effects are such as increasing effectiveness and efficiency of performance, productivity, and reliability of government accountability practices. For the negative effects, it also can be seen from the neglect of other quality of life standards, such as social capital, quality of life, equity, participation, social impact, community practices. All these issues tend not to get an essential place in the practices of local government accountability in Indonesia today.

4.2.4 Assumptions About Management and Managers (Leaders)

The fourth assumption of NPM implementation in the public sector is about the 'management culture' view: management is redefined as a separate organizational function, forms various new management positions, and emphasizes the authority of management and managers (leaders) above employees. Therefore, a new understanding of the manager (leader) appears. That is, the manager is quite responsible for managerial functions. This situation then has an impact on the method of managers (leaders) selection or appointment. In the selection phase, managers' abilities and managerial track record are more important. On the other hand, the manager's professional expertise and abilities in a case are less relevant to consider (Diefenbach, 2009; Kirkpatrick et al., 2005:3).

When this perspective is explored deeper, it is grounded from NPM ideology perspective itself, which tries to present a sense of profit-oriented organization in the realm of public sector organizations. Thus, the work of public sector organizations is considered as profit-oriented organizations work. Namely, work is neutral, rational, with a system that has been designed to change inputs into outputs with consideration of effectiveness and efficiency concerning the strategies that have been made previously (Diefenbach, 2009; Kärreman & Alvesson, 2004).

This perspective is then carried [consciously or unconsciously] in the practice of public sector management. As explained in the previous NPM assumptions, managers (leaders) tend to concentrate on the achievements they referred to their strategy and planning sheets. As happened also in Indonesia's case, the primary orientation of the manager is the achievement of the indicators, which are set out in the RPJMD and the performance agreement standards.

Furthermore, the perspective that managers or leaders selection / appointment are more considered by their managerial track records than their professional expertise and abilities in a working case (Diefenbach, 2009; Kirkpatrick et al., 2005:3) then opens the opportunity for political influence in the leaders selection, and efforts to reach the leadership chair (Diefenbach, 2005; Saunders, 2006). Primarily, when it is associated with Indonesian regional political context, it is familiar with the direct regional head elections held every five years. Then the opportunity is open. The elected regional leader *(Gubernur or Bupati)* will select managers of the regional agencies from the group of people who support him/her politically (Aspinall, 2013). Therefore, they can jointly prepare the agreed-upon performance indicators that are seen as easy to achieve together to secure a ticket to lead in the next period (Aspinall, 2013). On the other hand, someone who wants a position can also do political lobby so that he/she can be elected as manager (leader) of a regional agency (Aspinall, 2013; Diefenbach, 2005). Based on this explanation, it can be understood why strategic plans and accountability reports will look likely to be prepared with rhetoric and good things, as found by this study.

4.2.5 Assumptions About Employees and Work Culture

The fifth and final assumption regarding the application of NPM in the public sector is about employees and work culture. Diefenbach (2009) explains this assumption is based on the view that the public sector apparatus must develop a positive attitude like private sector employees at work. It is like developing a pro-active, collective, and entrepreneurial culture.

Theoretically, of course, there are many definite advantages of the proactive work culture of private companies' employees that can be followed and imitated by government officials. However, in reality, based on the result of researches in various countries in the world, the implementation of the NPM style in the public sector tends to lead to conditions of increased workload caused by the increasing number of routine administrative, supervisory, and audit obligations that must be carried out by public sector employees (Butterfield, Edwards, & Woodall, 2005; Kärreman & Alvesson, 2004; Saunders, 2006). This situation also has an impact of stress level increase experienced by public sector employees due to the amount of work they have to do (Kirkpatrick et al., 2005:176). Thus, the trend that occurs is a moral decreasing encouragement from employees in the public sector, in the form of a tendency to prioritize their self-interest. Employees then tend to pursue their performance indicators only. This situation then led to the loss of the substance of cooperation between colleagues and bureaus, which generally became the initial philosophy of the presence of the public sector's apparatus (Skålén, 2004).

In Indonesian context, this condition can also be seen in daily works of local government accountability reports preparation where local agencies are very focused on achieving the performance indicator points only. Thus, they are likely to ignore things that are not the primary reference of performance indicators, such as community protests over socio-environmental issues in the local government work area. This confirmed by results of this study and interviews with the LAKIP drafters of South Sulawesi government, as follows:

> And we also got data [for LAKIP compilation] that has been processed [by related government's agencies]...Those [related agencies] think the essential things are achieving their targets. Even if not achieved, but at least, it is close to [achieve the targets] ... (GOS1).

4.3 NPM Style: A Consequence of Neoliberal Developmentalism Paradigm

After seeing the influence of NPM, whether it is realized or not by the government officials, it is present in the praxis of Indonesian government. Furthermore, the important thing to study is the philosophy or the initial roots of NPM itself. It is because a good understanding of the philosophical basis of NPM can help us to criticize it to its foundation while being able to revise or to offer the antithesis of the application of NPM itself. Specifically, it is related to how NPM influences the praxis of Indonesian local government accountability, which is the focus of this book.

Referring to the experts, they generally agree to declare that NPM is a derivative even as it is a consequence of the neoliberalism adoption by a state or ruling regime (Connell, Fawcett, & Meagher, 2009; Diefenbach, 2009; Lorenz, 2012). In etymology, neoliberalism consists of two words, "neo" (Latin: new), and "liberalism" that related to a liberal economic ideology that was born in the nineteenth century. The economic liberalism believes that the freedom and recognition of individual rights, as well as free-market mechanisms (invisible hand), without state intervention [in term of regulation] is the primary driver of a country's economic growth (Bloom, 2017; Palley, 2005). In terminology, neoliberalism is understood as the rebirth of economic liberalism in this twentieth-twenty-first century (Bloom, 2017; Palley, 2005; Vincent, 2009).

The thing that distinguishes between liberal economic ideas of the nineteenth century and neoliberal in the twentieth-twenty-first century lies in the way both ideas view the function of the state. The understanding of economic liberalism in the nineteenth century alienated the role of the state or kingdom in regulating the market mechanism (Bloom, 2017; Palley, 2005). Meanwhile, neoliberalism transforms the role of the state in order to protect market mechanisms that occur by presenting regulations and policies that are pro-market interests (Boas & Gans-Morse, 2009; Wolff, 2012).

Many scientists believe that one consequence of neoliberalism is the globalization of today's economy. The globalization with all of its actors and structures starts from multinational corporations, world financial institutions (World Bank and IMF, WTO) and free trade and political agreements (such as; Washington consensus) on it (Fakih, 2001; Palley, 2005; Rodrik, 2012). As the result, countries tend to transform into corporatocracy, which must also compete with other countries to enhance their trade and markets, in the situation where natural resource supplies are decreasing and world population growth is increasing (Arthur, 2003; Peck & Tickell, 2002).

Based on these conditions, neoliberalism brought two main points that can be mapped as follows: first, with its free-market orientation, it has a consequence for removing all obstacles in the market. In particular, free from government interference, in other words, the government's policies must be pro-investment. For instance, the government should support investors by presenting the supporting infrastructure and recognize individual ownership by removing communal or customary ownership (Lloyd & Wolfe, 2016; Runk, 2012; Theodore, Peck, & Brenner, 2011). The government also should remove (lowering) taxes and not regulate prices in order to achieve a perfect competition in the market. Defenders of these neoliberalism policies believe that these policies will increase economic growth in a country (Springer, Birch, & MacLeavy, 2016). Second, neoliberalism tends to reduce government spending in the goods and services to encourage the growth of the private sector, which is expected to increase economic growth. One example is state subsidies revoke on important commodities such as education, health, food, housing, and energy. Therefore, more investors are present. Also, with perfect market competition, it will reach the optimum point of service and goods quality, which is ultimately expected to encourage economic growth (Jessop, 2012; Lorenz, 2012).

Of course, in applying this neoliberal perspective, a model of governance or management is needed to support the achievement of the neoliberalism's ultimate goal (Bevir, 2011). Moreover, it also requires the support of a [neo] liberal political regime as a necessity that market freedom is closely related to political freedom (Aspinall, 2013).

4.3.1 Harmony of Neoliberalism and NPM Values

Based on the explanation above, it can be seen that NPM is a governance model as a consequence of the [consciously or unconsciously] adoption of a neoliberalism viewpoint by a country. At least, there are several reasons why NPM can be said as a child of neoliberalism. As follows.

First, the consequences of neoliberalism's market-oriented view lead to the presence of the **NPM assumptions related to the business environment and strategic objectives of the government, which are also market-oriented and (investors) customer-oriented**. Governments using this NPM perspective, their policies and governance models tend to focus on bringing a pleasant investment climate. They will take several steps to support the good investment climate, such as focusing on encouraging the presence of investors from the private sector and presenting excellent infrastructures to support the business activities. By hoping, in the end, these steps will ultimately increase the country's economic growth (Lloyd & Wolfe, 2016; Runk, 2012; Theodore et al., 2011).

Second, the perspective of neoliberalism that believes by reducing government spending in the goods and services sector can encourage the growth of the private sector and increase the economic growth of a country. This view leads to the presence of **NPM assumptions related to management and performance indicators**. This perspective sees the performance of the government from its effective and efficient capabilities, specifically related to the budget expenditure and the time value of money. The more effective and efficient budget used by the government to complete a program, considered as a good thing (Bevir, 2011; Diefenbach, 2009). This perspective, consciously or not, builds a new perception in society that the government can not [optimally] implement a program if the budget is small or even zero. Naturally, this view will clash with the basic philosophy of the public government, which is service-oriented to the people, whatever its limitations (Skålén, 2004).

Third, by combining the two bases of the neoliberalism perspective, namely the market-oriented perspective and glorifying the reducing government budget based on an efficient and effective standard. This view then slowly creates hegemony on the perspective of the government apparatus. They become more focused to present **organizational structures and processes that are also effective and efficient**. This can be seen in countries that follow the NPM. The decentralization of government agencies becomes the first resort to maintain the effectiveness and efficiency of government apparatus work. This view also encourages the creation of a perfect competition climate between government and private institutions. Thus, the government agencies are forced to think about how to support themselves with a limited budget in order to operate optimally (Diefenbach, 2009; Lorenz, 2012).

Fourth, it comes with **the employee culture and management style that is in line with the perspective above**. As can be seen in various parts of the country implementing NPM; employees and managers in government agencies are likely to focus only on the effectiveness and efficiency of their performance indicators (Harun et al., 2013; Lee & Haque, 2006; Torres, 2004; Torsteinsen, 2012). In addition to that, a market-oriented perspective also dramatically influences the strengthening of political influence in the government's management. It is because the determination and assessment of performance indicators are closely related to the assessment of the ruling regime's political achievements (Aspinall, 2013; Diefenbach, 2005). By using this theoretical lens, it can also be understood why there are groups of people or issues that do not get a portion in policy, practice, to government accountability.

4.3.2 *Effects of Neoliberalism*

After the discussion about the harmony of the spirit of neoliberal economic understanding and NPM, this session will reveal what effects are likely to arise from the neoliberal regime. In the eyes of many critical thinkers on neoliberalism, it has many adverse effects. For example, neoliberalism strengthens the market's role in government (Boas & Gans-Morse, 2009; Wolff, 2012), cuts the government role from its main service sectors such as food supply, energy, education, and health (Baru & Mohan, 2018; Lorenz, 2012). Globalization then was born from the lack of market boundaries due to strengthening the role of the market (Rodrik, 2012). Neoliberalism and globalization also strengthen the financial market sector while increase the risk of financial crisis (Stein, 2012). It is also believed as the basis of a corporatocracy regime (Arthur, 2003; Peck & Tickell, 2002). Its market orientation also promotes the infrastructure-based development orientation to sustain market quality (Lloyd & Wolfe, 2016; Runk, 2012; Theodore et al., 2011). In the end, environmental damage took place due to the orientation of market interests above all (Hartwick & Peet, 2003; Harvey, 2007:173).

These situations also reflected in Indonesian context. Specifically, it is related to the policies orientation and practices of Indonesia's economic development, which tends to focus on development of supporting infrastructure for the sake of accelerated economic development and focused on the exploration of the natural resource sector. These orientations lead to the growth of socio-environmental adverse effects in the local regions, mostly, socio-economic conflicts between people versus corporations versus government that are related to the land, natural resources, and also resulted in environmental damage, as extensively advocated by local Non-Governmental Organizations (NGOs) (KPA, 2017a; LBH-Jakarta, 2015; WALHI, 2015, 2016).

From all the above explanations, the link between neoliberalism, NPM style, and its derivative effects can be described as follows (Fig. 4.2).

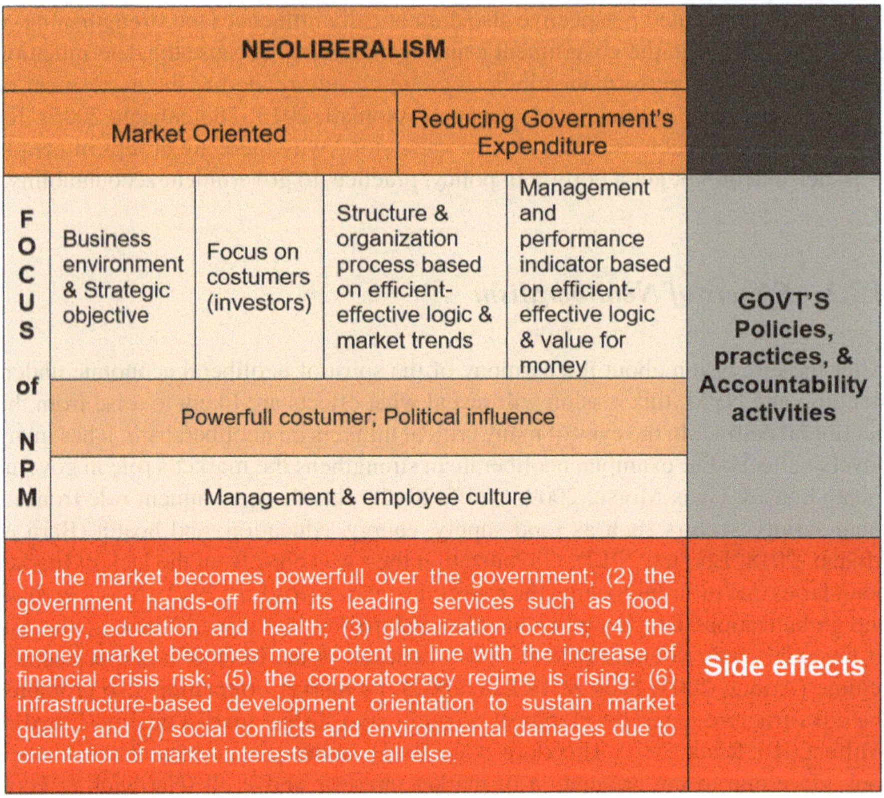

Fig. 4.2 Neoliberalism, NPM, and their Consequences

4.4 Summary of Criticism from NGOs on Government's Environmental and Social Governance Which Reflects Neoliberalism-NPM

As a result of the difficulty of accessing more study objects, I try to summarize NGOs' criticism of the government's governance and accountability related to environmental and social issues, which advocated by the leading socio-environmental NGOs in Indonesia. The primary source of information I used was the official annual report from the socio-environmental NGOs in the 2015–2019 period. Based on assumptions, their annual reports have an accumulation of official activity reports that can be verified for its quality of information. In addition, the annual report of socio-environmental NGOs is also appropriate to be used as a reference for the triangulation process of official information released by the government on social-environmental issues (Creswell, 2007:208).

I sampled three leading Indonesian NGOs, which actively advocate for social-environmental issues. Specifically, issues about conflicts in infrastructure and natural resource development in Indonesia related to the people's livelihoods, such as coastal reclamation, mining, and eviction of residents' land or houses for development reasons, as the focus of this research. The three socio-environmental NGOs are WALHI, KPA, and LBH.

4.4.1 Walhi

As the oldest socio-environmental NGO in Indonesia, WALHI has always been at the forefront of the struggle for the sustainability of ecology, and people's livelihoods advocacy in Indonesia. In particular, it is supported by its advocacy network throughout Indonesia (https://walhi.or.id/). Thus, reading its annual report termed "environmental outlook," the reader seems invited to look at the whole, or can be said to be a general summary, of the social-environmental issues that WALHI advocates throughout Indonesia. Furthermore, WALHI has consistently chosen the title "environmental outlook" by pairing the following year's number after the title for the annual report. For example, "2018 environmental outlook" which means the annual report for 2017 with forecasts to 2018. So, one can feel a sense of optimism for ecological sustainability from the beginning as he/she started reading the report.

In this study, WALHI annual reports used are based on the availability of such reports in the last five years (2015–2019) on the official website of WALHI (https://walhi.or.id/). The annual reports are: (1) 2015 environmental outlook: Collecting Promises, Demanding Change (WALHI, 2015); (2) 2016 environmental outlook: Requirements for Structural Reform for Improving Governance (WALHI, 2016); (3) 2018 environmental outlook: What is the Future of Ecological Justice 2018? (WALHI, 2018); (4) 2019 environmental outlook: Bringing Current Burden to the Future (WALHI, 2019).

When referring to the titles of those annual reports in general, the spirit of sustainable social-environmental governance can be felt to be a WALHI advocacy concern in the last five years (2015–2019). Furthermore, to obtain comprehensive information on how WALHI sees governance and accountability of Indonesian government (central to local) on socio-environmental issues, direct reading on the annual reports becomes essential.

In this research, the NVIVO-12 software helps the direct reading by researchers, primarily to facilitate the mapping and classification of important information raised by WALHI in its annual report. The following are results of the researchers' reading of the WALHI annual reports in the 2015–2019 period.

Table 4.3 Keywords found in WALHI's annual reports (2015–2019)

No.	2015		2016		2018		2019	
	Word	Q	Word	Q	Word	Q	Word	Q
1	Environment	190	Environment	80	Environment	157	Environment	396
2	Life	143	People	63	People	116	Life	331
3	WALHI	121	Forest	61	Life	115	WALHI	267
4	Forest	102	State	61	State	109	Politic	176
5	State	95	Life	52	Policy	105	Forest	146
6	Indonesia	81	Region	48	Corporation	80	State	113
7	Area	74	Policy	45	Politic	70	Indonesia	111
8	People	71	Permit	39	Development	63	Government	111
9	Region	66	Land	39	Indonesia	62	People	106
10	Development	58	Indonesia	37	Government	62	Disaster	105
11	Kalimantan	54	WALHI	36	Forest	60	Palm Oil	102
12	Conflict	54	Forestry	33	Investment	57	Investment	100
13	Sector	48	Development	33	Region	57	Conflict	75
14	Politic	47	Nature	28	Law	55	Area	74
15	Disaster	46	Disaster	28	WALHI	55	Law	73
16	Nature	45	Law	27	Agrarian	49	Policy	72
17	Mining	45	Area	27	Nature	44	People	69
18	Sumatera	42	Government	26	Area	44	Political party	69
19	Agrarian	41	Island	26	Sector	41	Nature	68
20	Policy	39	Conflict	25	Land	38	Right	60

It can be seen from the top 20 keywords found in WALHI's annual reports (2015–2019) that environmental issues in Indonesia tend to be related to government's issues, policy, politics, development, investment, people, conflict, and disaster. Even interestingly, it can be seen directly in each of these reports, WALHI emphasizes discussion on development and investment issues, as the root of environmental problems and social conflicts that arise in society.

Take for example, in the 2015 environmental outlook: Collecting Promises, Demanding Change (WALHI, 2015), it highlights the orientation of inviting as many investors as possible and building its supporting infrastructure as the root of natural resource conflicts and environmental damage in Indonesia as described by the organization in the chapter "Implications and Environmental impacts from the National and Global Policy" as follows:

> As a reminder, at the APEC Chief Executive Officer (CEO) meeting, Nusa Dua Bali on October 6, 2013, [President] SBY delivered an opening speech that vulgarly stated himself as a "sales" to attract investors who would invest in Indonesia. In his speech, SBY said, "To accelerate development, in May 2011, we launched the Master Plan for the Acceleration and Expansion of Indonesia's Economic Development (MP3EI) 2011–2025. Within the next 14 years, we are targeting 46 billion US $ to invest in 22 major economic activities, which

> is integrated in eight programs, which include mining, energy, industry, marine, tourism, and telecommunications". SBY, in his speech, also invited investors to enlarge business and investment opportunities in Indonesia ... At the APEC meeting in Bejing, the new government of Jokowi presented proposals that can be said to have no different from what SBY conveyed in the previous APEC forum. All available documents indicate Jokowi's proposal on infrastructure is another form of MP3EI that SBY had traded several years before. Infrastructure proposals that open the way for significant investments into Indonesia and with guarantees of land availability and mitigation of conflicts have the potential to seize community land and natural resources (WALHI, 2015:3).

Similar things were also found in other WALHI annual reports, as follows:

> In 2015, the economic policy package of the Jokowi-JK administration was quite surprising. It had a significant impact on the environment, namely deregulation policies in the natural resources sector, particularly forestry. Jokowi issued 5 Economic Packages, which are very friendly to the business world, "for the sake of" boosting the country's economic growth ... Prospective investors are given a quick service for granting investment licenses within three hours in the industrial area. By pocketing these permits, investors can immediately conduct investment activities. The regulations needed for this 3-h fast investment service are the Regulation of the Head of the Investment Coordinating Board (BKPM) and the Government Regulation on Industrial Estates and the Regulation of Minister of Finance...The Ministry of Environment and Forestry must cut the line of forestry sector permits ...Based on the timeline, investment permits must be cut from the previous 67 days to only 12 days. What truly worrying is how will the strategic environmental study and due diligence be fulfilled when the Regional Spatial Planning is not yet completed? .... If deregulation applies to the business world, this is not applied to local communities or indigenous peoples. The road is so long and complicated to gain recognition of the people's managed areas, for example, social forestry, which until now has not yet come out in the policy-law product. (WALHI, 2016:5)...National policy [later] made a significant contribution to perpetuating the conflicts of agrarian/natural resources and ecological, and until now, there has been no concrete agrarian conflict settlement agenda as promised by the President ... (WALHI, 2016:15).

Furthermore, looking at the criticism to Indonesian government raised by WALHI in its annual report, it can be seen neoliberal development paradigm features, which is highly market-oriented and investors-oriented, are inherent from a regime to a regime in Indonesia. In fact, in its 2018 annual report, WALHI concluded that the link between the massive exploitation of natural resources and the Indonesian political realm tended to be characterized by the corporatocracy state and [neo] liberal political schemes. As follows:

> 2018 as a political year is a year in which corporations begin to reinvest their political investment in politicians from potential regional leaders and to the members of national legislative candidates. It may also happen to the potential presidential candidates. Contributions through projects' fees implemented by prospective incumbents are also a normal form that will continue. For the state administrators that are seen uncooperative and hampering business, as usual, they may get some pressure, whether through formal legal systems, repositioning their strategic positions in government, and decay in the mass media. As well as, the people remain trapped in the transactional politics of liberal democracy. A sheet of [one hundred thousand rupiahs or more] becomes a tool of exchanging people's sovereignty with the political power of corrupt politicians, who are supported by the power of corporate capital (WALHI, 2019:41–42) ... The targeted economic growth of 5.2% to 5.6% still relies on investment growth. Infrastructure development is a key of investment service. The mode

> of development is still exploitative in natural resources. Although, there is a change in the focus of the sector. However, the metamorphosis still shows the direction of the capitalistic economy. Furthermore, the political momentum of 2018 is projected to have a risk for transactional politics to perpetuate the pace of liberal democracy (WALHI, 2019:50).

Above all the socio-environmental issues advocated by WALHI, it believes that one way to minimize the adverse effects of the government's perspective, which tends to focus on achieving market-oriented economic growth and investment-oriented, is to encourage transparent governance. Specifically, it relates to the transparency of business licenses in the natural resource sector that are vulnerable to cause conflicts with local communities and cause natural disasters. Hopefully, the transparency of information on natural resource management licenses will undoubtedly make it easier for the community to oversee the government's environmental management from the central to local levels and make it easier for the government to enforce environmental laws. As WALHI has advocated so far:

> … Encouraging transparency and the open system of natural resource licenses will be a part of breaking the tangled threads of the ecological conflict. WALHI's advocacy room is massive enough to use court instruments as a tool to demand ecological justice (WALHI, 2016:13).

4.4.2 KPA

In line with what was conveyed by WALHI in its annual report, KPA- Consortium for Agrarian Reform *(Konsorsium Pembaruan Agraria),* as an NGO focused on advocating of a fair agrarian system to the Indonesian people, specifically to farmers, fishermen and indigenous people-shows in its annual reports (2016–2019)[7] that this organization also focused on issues related to government, policy, politics, development, companies (investment), people (farmers and fishermen), and agrarian conflicts, as follows (Tables 4.4 and 4.5).

Furthermore, referring to the title of the KPA annual reports, such as (1) End-of-Year Note, 2016: Agrarian liberalization is intensified, agrarian reform is deflected; (2) End-of-Year Note, 2017: Agrarian Reform Under the Shadow of Investment; (3) End-of-Year Note, 2018: The Future of Agrarian Reform beyonds the Political Year; (4) Summary Report: Four Years of Implementation of Agrarian Reform, it can be seen that this NGO focus, during 2016–2019 period, was to advocate for agrarian problems with a focus on agrarian reform for the interests of the wider society.

Agrarian reform itself is a term about the process of restructuring the composition of ownership, control, and use of agrarian resources (especially land), for the benefit of the small people (small farmers, farm laborers, and the poor, as a whole and comprehensively Lee Peluso, Afiff, & Rachman, 2008; Wiradi, 2005:2). KPA makes it as its main campaign object because it believes that the starting point of all natual resource conflicts in various fields in Indonesia is based on problems with land

[7] these reports can directly be accessed through https://www.kpa.or.id/#.

Table 4.4 Keywords found in KPA's annual reports (2016–2018)

No	2016		2017		2018	
	Word	Q	Word	Q	Word	Q
1	Agrarian	128	Agrarian	378	Agrarian	492
2	Conflict	105	Conflict	262	Land	324
3	Land	72	Land	186	Conflict	253
4	People	64	Reformation	180	Reformation	220
5	Reformation	52	People	115	People	144
6	Government	45	Government	87	Forest	114
7	Plantation	34	Forest	75	Plantation	114
8	Forest	31	Area	67	Farmers	108
9	Rights	30	Ministry	62	Government	100
10	Area	28	Village	55	Politic	92
11	Sector	28	Sector	51	Private	84
12	Farmers	23	Plantation	48	Palm oil	78
13	Development	21	Development	47	Area	75
14	Region	21	Region	44	Sector	71
15	Settlement	19	Forestry	40	Rights	70
16	Policy	18	Policy	39	Article	69
17	Company	18	KPA	38	Company	69
18	KPA	17	National	37	KPA	62
19	Fishermen	17	Right	35	Renewal	61
20	Priority	17	Social	35	Policy	57

governance. Especially, policies on land ownership which are more oriented to the interests of companies and capital owners (investors). Yet, it unlikely to favor the poor people (KPA, 2016, 2017a, 2018, 2019).

> The high number of conflicts in the property sector (real estate, private industrial areas, etc.), is often closely related to infrastructure development projects ... Infrastructure development and integrated regional development are needed to spur the wheels of a country's economy. However, unilateral and coercive, non-participatory, non-transparent, corrupt, manipulative, business orientation are stronger than the public interest, up to a repressive approach as a response to the people's resistance, always decorating land acquisition and acquisition projects for infrastructure development and property. This situation is contrary to the discourse that was built, which development aims to boost the economy and welfare of the people (KPA, 2018: 24).

The spirit of agrarian reform is actually in line with the focus of the Indonesian government in the 2014–2019 period (Presidential Regulation of the Republic of Indonesia Number 2/2015 about the National Medium-Term Development Plan 2015–2019 pages: 6). However, in practice in the field, as found by this NGO, conflict over natural resources is still quite high in number and extent of its impact from year to year. As recorded in KPA Annual Report.

Table 4.5 Agrarian conflicts in Indonesia (KPA's Annual Report 2016–2018)

	Year		
Conflict in	2016	2017	2018
Plantation	163	208	144
Agriculture	7	78	53
Forestry	25	30	19
Property	117	199	137
Infrastructure	100	94	16
Coastal/marine	10	28	12
Mining	21	22	29
Oil and gas	7	–	–
Quantity	**450**	**659**	**410**
Impact			
Total Area (Hectare)	**1.265.027**	**520.491,87**	**807.177,613**
Total Household	**86.745**	**652.738**	**87.568**

From the sectors that reported by this organnization, it can be seen that there is a relationship between agrarian conflicts with the leading sectors of Indonesia's economic drivers. Thus, as mentioned earlier, the government policies that tend to focus on business interests (investors) compared to the public interest can be said as the root of agrarian conflicts themselves (KPA, 2018:24). Furthermore, this also signifies the characteristic of neoliberalism, which prioritizes the interests of the market over the interests of the people at large. Realize it or not, it has become a general policy of the Indonesian government.

Above all, KPA believes that besides ending the paradigm of capitalistic agrarian political economy, which prioritizes the interests of investors (KPA, 2018:2), the important thing that should be encouraged is the presence of transparency from the government to the public regarding the information on concessions of state and private plantation companies (HGU,[8] location permits, etc.) (KPA, 2019:6). This transparency can be a way to prevent and resolve conflicts and reduce the inequality that has occurred because the government and community have the same reference to oversee corporate business practices in sectors that vulnerable to agrarian conflicts (KPA, 2019:6). This is also in line with the legitimacy theory, which believes the presence of regulation, and public oversight on the business practices could encourage businesses to act following what is expected by the community (Archel, Husillos, Larrinaga, & Spence, 2009; Ieng Chu, Chatterjee, & Brown, 2012; Lanis & Richardson, 2013).

[8]In Bahasa Indonesia: *HGU: Hak Guna Usaha.* It is based on Article 28 paragraph (1) of the Basic Agrarian Law Number 5/1960 *(UUPA),* HGU is the right to cultivate land that is directly controlled by the state within a certain period for agricultural, fisheries, or livestock companies.

4.4.3 LBH

Furthermore, it is interesting to see the context of social-environmental governance and conflicts that arise in the urban context. In this case, the report issued by LBH: Legal Aid Foundation (Lembaga Bantuan Hukum-Jakarta)–as an institution that focuses to provide legal assistance to people in need– can be a reference. From 2015 to 2018, LBH-Jakarta issued an annual research report on forced evictions that occurred in the city of Jakarta, the capital of Indonesia.

The reports are interesting because they focus to document and advocate for forced eviction acts, which are carried out by the government to the people [mostly poor people] in the city of Jakarta. These reports are in line with the primary objective of this book, which is to present the emancipatory socio-environmental accounting idea from NGOs advocacy activities that concern on socio-environmental issues that often do not get a portion in government policies, practices, and accountability. The LBH-Jakarta research reports can then be presented as a mirror to measure how the local government is responsible for eviction policies, especially by referring to the universal human rights standards that are also ratified by Indonesia. As said by this NGO in its report:

> The development that took place in DKI Jakarta, the capital of Indonesia, tended to violate the fundamental rights of victims of forced evictions. Even though Indonesia has ratified the International Covenant on Economic, Social, and Cultural Rights since 2005. Article 11 of the Covenant guarantees the right of everyone to obtain adequate housing. Furthermore, the General Opinion of the Committee on Economic, Social, and Cultural Rights ("CESCR") No. 7 year 1997, stated that eviction could affect the right to life, right to security, right not to interfere with his personal, family, and household affairs, as well as rights to guarantee of ownership. The UN Human Rights Commission categorizes forced evictions as a form of violations of human rights through resolution number 2008/24 (LBH-Jakarta, 2015:1) ...This report was made to document the forced evictions that occurred in DKI Jakarta throughout 2015. It is expected that information obtained through this research can be used to advocate the government policies, especially the policies of Provincial Government of DKI Jakarta, which is planned to conduct forced evictions of many citizens in 2016 (Kompas, 2015). It is hoped that the government will use the perspective of human rights in implementing the eviction policies (LBH- Jakarta, 2015:2).

Furthermore, referring to the annual research reports of this organization, evictions were generally carried out by the city government based on orientation to restrain *(penertiban)* (as the most reason), normalization projects, revitalization of areas, city parks, MRT projects, road widening, and the green belt (LBH-Jakarta, 2015:19, 2017a:32, b:34, 2018:22). However, in the opinion of LBH-Jakarta (2017b:5–6) the meaning of "restrain" based on the provisions of regional regulations (DKI Jakarta Regional Regulation No. 8/2007 on Public Order)[9] is very flexible. It opens broad interpretations because there are no strict restrictions on forms of "restrain" actions so that the term "restrain" can be categorized as political terminology, not technical terminology. This situation then opens various interpretations that can differ depending on who is in power. In addition, the term "restrain" tends to be a subtle

---

[9] *Peraturan Daerah DKI Jakarta Nomor 8 Tahun 2007 tentang Ketertiban Umum.*

choice of words to describe the forced evictions for the development of cities (Audefroy, 1994; LBH-Jakarta, 2015:1). Thus, LBH-Jakarta (2017b:5–6) assesses that the government's "restrain" action to the community can be categorized as an act of forced eviction if it is against human rights standards (LBH-Jakarta, 2015:1).

Moreover, when referring to cases of forced evictions reported by the NGO (see: Table 4.6), it can be seen that in general, the procedure of forced evictions is carried out without prior consultation to find a win–win solution between the government and affected communities. This condition indirectly violates the human rights following the general provisions of the Committee on Economic, Social, and Cultural Rights ("CESCR") Number 7/1997 and Law Number 11/2005[10] (LBH-Jakarta, 2015:5,8). LBH-Jakarta also adds, especially if it is associated with the use of state apparatus, violence in practice, and does not provide compensation solutions following the rights to a decent life. It can be said that the government's "restrain" approach (read: forced evictions) is far from complying with the standards of Universal Human Rights (LBH-Jakarta, 2015, 2017a, b, 2018).

Table 4.6 Forced evictions in Jakarta (LBH-Jakarta's Annual Report 2015–2018)

	2015	2016	2017	2018
Total household location	–	90	20	17
Total business location	–	97	82	53
Total household + business location		6	8	9
Total number of forced eviction location	113	193	110	79
Victims				
Total household (HH)	8.145 HH	5.726 HH	1070 HH	186 HH
Total business unit (BU)	6.283 BU	5.379 BU	1.642 BU	773 BU
Total people's home (HH) + business unit (BU)	–	–	112 HH + 89 BU	89 HH + 93 BU
Procedure of eviction				
one-sided-from the developer (Government/Private parties)	95	68[a]	79	64
Negotiation: between developer & affected people	18	5[a]	5	7
People volunteery	–	–	–	–
No information	–	23[a]	26	8

Notes
[a]These informations are based on Total Household location + (Total Household + Business location) in 2016

[10]Indonesia has ratified the International Covenant on Economic, Social and Cultural Rights (International Covenant on Economic Social and Cultural Rights), hereinafter referred to as ICESCR, with Law Number 11/2005.

The exciting thing, LBH-Jakarta, emphasized the government should use a participatory approach to avoid human rights violations in development by inviting citizens to jointly design the solutions to the urban issues (LBH-Jakarta, 2015, 2017a, b, 2018). In addition, the emphasis on transparency of information on city policing activities, related to the objectives and parties involved in policing, becomes essential to be accessible by the community. Thus, the affected citizens can demand protection and accountability to the parties who responsible, as well as this is also under UN's universal human rights provisions[11] (LBH-Jakarta, 2017b:23).

These two important points from LBH-Jakarta generally illustrate that they are in line with what WALHI and KPA struggling for to unravel the tangle of the development paradigm impact. In other words, it can be said that in general, the leading NGOs in Indonesia have a voice to support transparent, sustainable, and participatory development policies and practices, which promote ecological justice, and are friendlier to human rights.

References

Aljazeera.com. (2019). Environment a missing topic as Indonesia election looms. Retrieved August 2, 2019, from https://www.aljazeera.com/news/2019/04/environment-missing-topic-indonesia-election-looms-190408080355562.html.

Archel, P., Husillos, J., Larrinaga, C., & Spence, C. (2009). Social disclosure, legitimacy theory and the role of the state. *Accounting, Auditing & Accountability Journal, 22*(8), 1284–1307. https://doi.org/10.1108/09513570910999319.

Arthur, M. (2003). *Struggle and the prospects for world government.* Trafford Publishing.

Aspinall, E. (2013). A nation in fragments: Patronage and neoliberalism in contemporary Indonesia. *Critical Asian Studies, 45*(1), 27–54. https://doi.org/10.1080/14672715.2013.758820.

Audefroy, J. (1994). Eviction trends worldwide and the role of local authorities the right to housing. *Environment and Urbanization, 6*(1), 8–24. https://doi.org/10.1177/095624789400600102.

Baru, R. V., & Mohan, M. (2018). Globalisation and neoliberalism as structural drivers of health inequities. *Health Research Policy and Systems, 16*(1), 91. https://doi.org/10.1186/s12961-018-0365-2.

Belal, A. R., Cooper, S. M., & Khan, N. A. (2015). Corporate environmental responsibility and accountability: What chance in vulnerable Bangladesh? *Critical Perspectives on Accounting, 33*, 44–58. https://doi.org/10.1016/j.cpa.2015.01.005.

Bevir, M. (2011). Governance and governmentality after neoliberalism. *Policy and Politics, 39*(4), 457–471. https://doi.org/10.1332/030557310X550141.

Bloom, P. (2017). *The ethics of neoliberalism: The business of making capitalism moral.* Routledge.

Boas, T. C., & Gans, J. (2009). Neoliberalism: From new liberal philosophy to anti-liberal slogan. *Studies in Comparative International Development, 44*(2), 137–161.

Butterfield, R., Edwards, C., & Woodall, J. (2005). The new public management and managerial roles: The case of the police sergeant. *British Journal of Management, 16*(4), 329–341. https://doi.org/10.1111/j.1467-8551.2005.00466.x.

Chandra, W. (2016). Tolak Reklamasi Pantai Losari, Walhi Gugat Gubernur Sulsel ke PTUN. Retrieved September 19, 2019, from Mongabay website: https://www.mongabay.co.id/2016/02/14/tolak-reklamasi-pantai-losari-walhi-gugat-gubernur-sulsel-ke-ptun/.

[11]CESCR Number 7 year 1997 about Forced Evictions and the United Nations Basic Principles and Guidelines on Development-Based Evictions.

Connell, R., Fawcett, B., & Meagher, G. (2009). Neoliberalism, new public management and the human service professions: Introduction to the special issue. *Journal of Sociology, 45*(4), 331–338. https://doi.org/10.1177/1440783309346472.

Courpasson, D. (2000). Managerial strategies of domination. Power in soft bureaucracies. *Organization Studies, 21*(1), 141–161. https://doi.org/https://doi.org/10.1177/0170840600211001.

Creswell, J. W. (2007). *Qualitative inquiry & research design choosing among five approaches*. London: Sage Publications Inc.

Deegan, C. (2002). Introduction: The legitimising effect of social and environmental disclosures–a theoretical foundation. *Accounting, Auditing & Accountability Journal, 15*(3), 282–311.

Diefenbach, T. (2004). Different meanings of intangible assets and knowledge: and their implications for management and innovation. *International Journal of Knowledge, Culture and Change Management, 4*.

Diefenbach, T. (2005). Competing strategic perspectives and sense-making of senior managers in Academia. *The International Journal of Knowledge, Culture, and Change Management: Annual Review, 5*(6), 126–137. https://doi.org/https://doi.org/10.18848/1447-9524/cgp/v05i06/49910.

Diefenbach, T. (2007). The managerialistic ideology of organisational change managment. *Journal of Organizational Change Management, 20*(1), 126–144. https://doi.org/10.1108/095348107107 15324.

Diefenbach, T. (2009). New public management in public sector organizations: The dark sides of managerialistic "Enlightenment." *Public Administration, 87*(4), 892–909. https://doi.org/10.1111/j.1467-9299.2009.01766.x.

Environmental Justice Atlas. (2019). Takalar Fisherfolk against Sea Sand Mining, Sulawesi, Indonesia. Retrieved December 20, 2019, from https://ejatlas.org/conflict/takalar-fisherfolk-aga inst-sea-sand-mining.

Fakih, M. (2001). *Runtuhnya Teori Pembangunan and Globalisasi*. Yogyakarta: InsistPress.

Hartwick, E., & Peet, R. (2003). Neoliberalism and nature: The case of the WTO. *The Annals of the American Academy of Political and Social Science, 590*(1), 188–211.

Harun, H., An, Y., & Kahar, A. (2013). Implementation and challenges of introducing NPM and accrual accounting in Indonesian local government. *Public Money and Management, 33*(5), 383–388. https://doi.org/10.1080/09540962.2013.817131.

Harvey, D. (2007). *A brief history of neoliberalism*. USA: Oxford University Press.

Hope, K. R., Sr. (2001). The new public management: Context and practice in Africa. *International Public Management Journal, 4*(2), 119–134.

Ieng, C., Chatterjee, B., & Brown, A. (2012). The current status of greenhouse gas reporting by Chinese companies: A test of legitimacy theory. *Managerial Auditing Journal, 28*(2), 114–139.

Jessop, B. (2012). Liberalism, neoliberalism, and urban governance: A state-theoretical perspective. *Spaces of Neoliberalism: Urban Restructuring in North America and Western Europe,* 104–125. https://doi.org/10.1002/9781444397499.ch5.

Karp, T. (2005). Unpacking the mysteries of change: Mental modelling. *Journal of Change Management, 5*(1), 87–96.

Kärreman, D., & Alvesson, M. (2004). Cages in Tandem: Management control, social identity, and identification in a knowledge-intensive firm. *Organization, 11*(1), 149–175. https://doi.org/10.1177/1350508404039662.

Kirkpatrick, I., Ackroyd, S., & Walker, R. (2005). *The new managerialism and public service professions*. In Hampshire: Palgrave Macmillan. Springer.

KPA. (2016). *Catatan Akhir Tahun 2016: Liberalisasi agraria diperhebat, reforma agraria dibelokkan*. https://doi.org/https://doi.org/10.1017/CBO9781107415324.004.

KPA. (2017a). *Catatan Akhir Tahun 2017 Reforma Agraria Di Bawah Bayangan Investasi*. Retrieved from https://kpa.or.id/publikasi/baca/laporan/25/Catahu_2017:__Reforma_Agraria_Dibawah_B ayang_Investasi__Gaung_Besar_Dipinggiran_Jalan/.

KPA. (2018). *Catatan Akhir Tahun 2018: Masa Depan Reforma Agraria Melampaui Tahun Politik*. Retrieved from https://kpa.or.id/publikasi/baca/laporan/30/Catahu_2018:_Masa_Depan_Refo rma_Agraria_Melampaui_Tahun_Politik/?lang=id.

KPA. (2019). *Laporan Ringkas: Empat Tahun Implementasi Reforma Agraria.* Retrieved from https://kpa.or.id/publikasi/baca/laporan/77/Empat_Tahun_Implementasi_Reforma_Agraria/?lan g=id.

Lanis, R., & Richardson, G. (2013). Corporate social responsibility and tax aggresiveness: A test of legitimacy theory. *Accounting, Auditing & Accountability Journal, 26*(1), 75–100.

LBH-Jakarta. (2015). *Atas Nama Pembangunan: Laporan Penggusuran Paksa di Jakarta Tahun 2015* (Vol. 2017). Retrieved from https://www.bantuanhukum.or.id/web/atas-nama-pemban gunan-laporan-penggusuran-paksa-di-jakarta-tahun-201.

LBH-Jakarta. (2017a). *Mengais di Pusaran Janji: Laporan Penggusuran paksa di Wilayah DKI Jakarta tahun 2017*. Retrieved from https://www.bantuanhukum.or.id/web/mengais-di-pusaran-janji-laporan-penggusuran-paksa-di-wilayah-dki-jakarta-2017/.

LBH-Jakarta. (2017b). *Seperti Puing: Laporan Penggusuran Paksa di Wilayah DKI Jakarta Tahun 2016*. Retrieved from https://www.bantuanhukum.or.id/web/wp-content/uploads/2017/04/ Laporan-Penggusuran_LBHJAKARTA_2016.pdf.

LBH-Jakarta. (2018). *Masih Ada: Laporan Penggusuran Paksa di Wilayah DKI Jakarta Januari-September 2018*. Retrieved from https://www.bantuanhukum.or.id/web/masih-ada-laporan-pen ggusuran-paksa-di-wilayah-dki-jakarta-2018/.

Lee , N., Afiff, S., & Rachman, N. F. (2008). Claiming the grounds for reform: Agrarian and environmental movements in Indonesia. *Journal of Agrarian Change, 8*(2–3), 377–407.

Lee, E. W. Y., & Haque, M. S. (2006). The new public management reform and governance in Asian NICs: A comparison of Hong Kong and Singapore. *Governance, 19*(4), 605–626. https://doi.org/ 10.1111/j.1468-0491.2006.00330.x.

Lloyd, D., & Wolfe, P. (2016). Settler colonial logics and the neoliberal regime. *Settler Colonial Studies, 6*(2), 109–118. https://doi.org/10.1080/2201473X.2015.1035361.

Lorenz, C. (2012). If you're so smart, why are you under surveillance? Universities, neoliberalism, and new public management. *Critical Inquiry, 38*(3), 599–629. https://doi.org/10.1086/664553.

Michael, B. (2005). Questioning Public Sector Accountability. *Public Integrity, 7*(2), 95–109. https://doi.org/10.1080/10999922.2005.11051275.

Mongabay. (2017a). Aksi Warga Takalar Menolak Tambang Pasir: Jangan Paksa Kami Menjadi Teroris. Retrieved August 1, 2019, from https://www.mongabay.co.id/2017/06/17/aksi-warga-takalar-menolak-tambang-pasir-jangan-paksa-kami-menjadi-teroris/.

Mongabay. (2017b). Sand mining, land reclamation meet fierce resistance in Makassar. Retrieved March 10, 2019, from https://news.mongabay.com/2017/07/sand-mining-land-reclam ation-meet-fierce-resistance-in-makassar/.

Mongabay. (2020). Tambang Pasir Laut Galesong Rusak Ekosistem Laut dan Sebabkan Abrasi. Retrieved January 29, 2020, from https://www.mongabay.co.id/2020/01/29/tambang-pasir-laut-galesong-rusak-ekosistem-laut-dan-sebabkan-abrasi/.

Palley, T. I. (2005). From Keynesianism to neoliberalism: Shifting paradigms in economics. *Neoliberalism: A Critical Reader*, (April), 20–29.

Peck, J., & Tickell, A. (2002). Neoliberalizing space. *Antipode, 34*(3), 380–404. https://doi.org/10. 1111/1467-8330.00247.

Pina, V., & Torres, L. (2003). Reshaping public sector accounting: An international comparative view. *Canadian Journal of Administrative Sciences/Revue Canadienne Des Sciences De l'Administration, 20*(4), 334–350.

Rodrik, D. (2012). *The globalization paradox: Democracy and the future of the world economy.* Washington: Harvard Uiversity Press.

Runk, J. V. (2012). Indigenous land and environmental conflicts in Panama: neoliberal multiculturalism, changing legislation, and human rights. *Journal of Latin American Geography*, 21–47.

Ruslin, I. T. (2017). Subaltern Dan Kebijakan Pembangunan Reklamasi Pantai Di Kota Makassar. *JPP (Jurnal Politik Profetik), 5*(2), 185–199.

Saunders, M. (2006). The madness and malady of managerialism. *Quadrant, 50*(3), 9.

Skålén, P. (2004). New public management reform and the construction of organizational identities. *International Journal of Public Sector Management, 17*(3), 251–263.

Springer, S., Birch, K., & MacLeavy, J. (2016). *Handbook of neoliberalism.* Routledge.

Stein, H. (2012). The neoliberal policy paradigm and the great recession. *Panoeconomicus, 59*(4), 421–440.

SUARA.COM. (2018). Ini Alasan Warga Sekitar Selamanya Tolak Bandara Baru Yogyakarta. Retrieved August 2, 2019, from https://www.suara.com/bisnis/2018/02/08/130241/ini-alasan-warga-sekitar-selamanya-tolak-bandara-baru-yogyakarta.

Theodore, N., Peck, J., & Brenner, N. (2011). Neoliberal urbanism: cities and the rule of markets. *The New Blackwell Companion to the City, 1625.*

Torres, L. (2004). Trajectories in public administration reforms in European Continental countries. *Australian Journal of Public Administration, 63*(3), 99–112. https://doi.org/10.1111/j.1467-8500.2004.00394.x.

Torsteinsen, H. (2012). Why does Post-Bureaucracy lead to more formalisation? *Local Government Studies, 38*(3), 321–344. https://doi.org/10.1080/03003930.2011.629194.

Vincent, A. (2009). *Modern political ideologies.* Wiley.

WALHI. (2015). *Tinjauan Lingkungan Hidup 2015 WALHI.* Retrieved from https://walhi.or.id/wp-content/uploads/2016/12/OutLook2015_W.pdf.

WALHI. (2016). *Tinjauan Lingkungan Hidup WALHI 2016: Keharusan Pembenahan Struktural untuk Perbaikan Tata Kelola.* Retrieved from Wahana Lingkungan Hidup Indonesia website: https://www.walhi.or.id/2016/12/22/outlook-2016-tujuan-lingkungan-hidup/.

WALHI. (2018). *Tinjauan Lingkungan Hidup 2018 : Bagaimana Masa Depan Keadilan Ekologis 2018*? Retrieved from https://walhi.or.id/wp-content/uploads/2018/12/Layout_Tinjauan-Lingkungan-2018.pdf.

WALHI. (2019). *Tinjauan Lingkungan Hidup 2019: Membawa Beban Masa Kini ke Masa Depan*? Retrieved from https://walhi.or.id/tinjauan-lingkungan-hidup-2019/.

Wiradi, G. (2005). *Reforma Agraria untuk Pemula.* Retrieved from https://kpa.or.id/publikasi/download/74032-reforma-agraria-untuk-pemula.pdf.

Wolff, R. D. (2012). *Democracy at work: A cure for capitalism.* Haymarket books.

Chapter 5
Discussions: Interpretation and Analysis

If the system cannot be verified and cannot be criticized,
Then the science is dead.
—Tan Malaka (Indonesian's National Hero and intellectual)

The previous chapter described the initial findings on the issue of local government socio-environmental accountability that examined in this study. Therefore, in this chapter, the discussion will focus to analyze and interpret this study findings using Tan Malaka's critical methodology, which was developed specifically for this research. Hopefully, it will achieve a new synthesis to develop the regional government accountability framework that integrates and emancipates social-environmental issues often faced by poor people and advocated by social-environmental NGOs in Indonesia. These social-environmental issues so far tend to be marginalized by market interests and economic development issues in the government's policies to accountability practices. Other expectations, the results of this research analysis will further support and encourage the achievement of sustainable development goals that are becoming a worldwide concern today.

Furthermore, there are three main focuses and values of the critical methodology of Tan Malaka developed and integrated into the steps of this study (see again: Chapter three). These values also become the basis of this study results analysis. The values are: (1) Tan Malaka's view emphasizes the rejection to direct adoption of "Western (foreign)" concept in Indonesian government management system. In line with this research context, it is related to criticize the trends in government accountability systems that tend to adopt NPM styles as a consequence of neoliberalism (Diefenbach, 2009; Harun, An, & Kahar, 2013). (2) Tan Malaka's view stresses the importance of every community class and group to be represented so they have a place in the building of a government management system. Therefore, the government management system should be oriented to public interests, which is also following the philosophical foundation of Indonesia (Pancasila) (Latif, 2011). (3) Tan Malaka's thought focuses on how the social structures and agents (individual/society) can influence one another. Thus, policies and regulations can be based on the societies'

H. M. Shahib, *Towards the Local Government's Integrated Accountability Framework*,
https://doi.org/10.1007/978-981-33-6617-6_5

interests and be the basis to oversee the government and industry performance for the sake of the people. The explanation in this chapter will be in harmony with three main focuses of Tan Malaka's critical values.

5.1 Synthesis of Research Results: Emphasizes the Rejection of Direct Adoption of "Western (Foreign)" Concept in the Indonesian Government Management System

The initial synthesis of this study results will be based on the first value of Tan Malaka's critical methodology that developed for this research. It emphasizes the rejection of "Western (foreign)" concept direct adoption in the Indonesian government management system. This value is one of the main essences of Tan Malaka's view in his book "Parliament or Soviet?" (Malaka, 1921) which explains the connection between the state governance and capitalism in influencing the governance style in a country. Therefore, it tends to lead to the presence of policies that harm people. Based on this critical reading, Tan Malaka has warned Indonesians since a long time ago that the adoption of [foreign] western's model management style does not guarantee it to be suitable and sufficient to be applied in Indonesian context (Malaka, 1921). As he said:

> "Throughout our knowledge, it is enough to show the nature, origin, and institutions of the West, one of which we must choose if we have the freedom to govern our own country (Malaka, 1921:42)……All domestic institutions without nationality would undoubtedly corrupt the nation (Malaka, 1921:43)."

When this perspective is reflected in today's conditions, specifically related to the local government accountability system that is the focus of this study, in my opinion, it is still very coherent. Moreover, the perspective of Tan Malaka encourages the public to use their ability to properly understand characteristics and origins of western (foreign) institutions or systems. Then the public should compare them to the needs (believed values) of Indonesian people before using them in a government system.

With this perspective, the implementation of Indonesian government's accountability system can also be criticized in order to optimize its benefits for Indonesians. In addition to that, as discussed previously in chapter four, the system of Indonesian local governments accountability tends to marginalize the values that are believed by the people, specifically related to how people are responding to social and environmental issues that affect their lives due to the (neoliberalism) NPM model, consciously or unconsciously, has so dominating the mindset and practice of Indonesian government governance.

Furthermore, I would like to go back a little to the discussion of this critical methodology values in chapter three, which is in harmony with the spirit of reflexivity that is encouraged to be used in qualitative research (Mason, 2002:149; Piantanida &

Garman, 2009:63). I argue that the first value of Tan Malaka's critical methodology, when it is contextualized with this study's objects will give rise to the three main values, namely: (1) criticizes the forms of NPM implementation in local government accountability systems that do not accommodate the people's aspirations and values, especially those affected by social and environmental problems. Then, as Tan Malaka stressed the importance of nationalism as the spirit of a state institution, the second value was translated by (2) promoting the values and voices of people who are not accommodated in the government's accountability system, to be given a proper space in the government's accountability system. (3) In the end, it will lead to the presence of more social-environmental values and indicators in the government accountability system that based on the people's local principles and aspirations. Based on these values, in this section I will reveal and synthesize the findings of this research under the value of Tan Malaka's critical methodology.

5.1.1 Critics to NPM's Styles in Local Government Accountability Systems

As described in Chapter Four, this study found exciting facts related to how NPM style was present in the local government accountability system (SAKIP) to be presented in its final report (LAKIP). By understanding NPM patterns, it will be beneficial to understand why aspirations and values embraced by the community affected by social-environmental issues are not being accommodated in the system to government accountability reports.

According to this study findings, the pattern of NPM in the government accountability system can be explained by following the logic of government regulations-policies and practices. As follows.

5.1.1.1 From the Regulation Side

According to history, the presence of a report-based government performance accountability system cannot be separated from changes in the political-economic structure in Indonesia after the 1997–1998 economic crisis and the 1998-reform movement. Mainly, it related to the agenda of bureaucratic reform to accelerate the improvement of the country after the 1998's economic and political crisis (Harun et al., 2013).

Interestingly, the agenda to improve the bureaucracy is also one of the requirements to receive loan assistance from IMF, which is stated in the Letter of Intent between Indonesian government and IMF. Specifically, it is related to strengthen the public expenditure management point. One of its derivatives is to improve accountability in managing the budget to focus on development. This is done at the same time with shrinking the role of the state in the market, by reducing energy (oil and gas) subsidies and promoting market and BUMN (State-Owned Enterprises) privatization. From this point, several scholars consider that there is a close relationship between the neoliberal spirit -which is highly market-oriented, economic development-oriented, and reducing the government spending-oriented-brought by the IMF with the Indonesian government's development orientation (Ramli & Nuryadin, 2007; Robison & Rosser, 1998; Rosser, 2013).

The conditions then slowly [consciously or not] instill the logic of NPM-as a derivative of neoliberalism—for example, focused on efficient-effective standards and budget achievement (value for money)—in government's policies, practices, and accountability. Furthermore, the 90s were the time when many countries in the world began to adopt NPM as these countries believe NPM is the right tool to reform the country's administration and support economic growth (Diefenbach, 2009; Kirkpatrick, Ackroyd, & Walker, 2005:13; Lee & Haque, 2006; Torres, 2004). Notably, the NPM discourse was supported by many researchers and international institutions. Therefore, NPM was increasingly widely accepted by international community (Diefenbach, 2009; Kirkpatrick et al., 2005:13; Lee & Haque, 2006; Torres, 2004), including Indonesian government (Harun et al., 2013).

The characteristics of NPM can then be seen from government regulations related to the government accountability system that was born later. From the first regulation Presidential Instruction Number 7/1999 about the Performance Accountability of Government Agencies to the newest, such as The President Regulation Number 29/2014 as a legal basis of the system of performance accountability in government agencies *(SAKIP)*,[1] and the Regulation of the Minister of State-Apparatus Employment *(Permenpan)* Number 29/2010 to act as a guideline to prepare the accountability reports as an integral part of the government accountability system. All those regulations standardize the orientation of effective and efficient government practices in the accountability system, with emphasis in harmony with the administration of the Government's Budget and Accounting System, control activities, and evaluation of the development plan implementation.

Interestingly, when referring to Tan Malaka's view related to the importance of making criticisms and adjustments to the western (foreign) system before using it for the benefit of the people, Indonesian government has also made these adjustments. As stated by this research informant as a representative of the standard setter and supervisor of Government's Accountability System[2]:

[1] *Sistem Akuntabilitas Kinerja Instansi Pemerintah (SAKIP).*

[2] Ministry of Administrative and Bureaucratic Reform (KEMENPAN-RB) as the standard setter and supervisor of Government's Accountability System in Indonesia.

> So, it was our policy to have SAKIP because we wanted to follow GPRA-USA[3]... Then it developed [adapted to the needs] and then came the name of strategic management, strategic planning, now called SAKIP since 1999[4] (MA1).

In addition, the adjustment of the adoption of accountability systems to the needs of Indonesians, in particular, is reflected in the presence of The Government Regulation of the Republic of Indonesia Number 8/2008 about Steps, Procedures for Preparation, Control and Evaluation of Regional Development Plan Implementation.[5] This regulation is also the basis that binds the development plan, policy implementation, and accountability of government performance from upstream to downstream. In this regulation is clearly written that local government obligations to hold *Musrenbang*,[6] the inter-stakeholders[7] forum in the context of preparing regional development plans before the RPJMD[8] ratified as the local government's regulation, and become the standards for governments' accountability performance indicators (see again Fig. 4.1 in 4.1 section).

The Government Regulation of the Republic of Indonesia Number 8/2008 supports bottom-up participatory development patterns by opening space for a direct community involvement in a sustainable development planning. By expecting, what is expected by the people or what is the community's concern about development in the region can be a program that is considered by the government as stated in the regulation:

> Regional development planning is formulated in a transparent, responsive, efficient, effective, accountable, participatory, measurable, fair, and sustainable manner (Ch.II Principles of Regional Development Planning article 3) ...Musrenbang is an inter-stakeholders forum in the context of formulating regional development plans ... Stakeholders are those who directly or indirectly benefited from the impact and planning of regional development ... (Ch. I general provisions of Article 1).

Even interestingly, on RPJMD[9] preparation, in fact, to reassure the presentation of participatory and environmentally planning, the government has also presented

---

[3]GPRA-USA (Government Performance and Results Act of 1993 is a United States law enacted in 1993). It is designed to improve government performance management. By requiring institutions to engage in performance management tasks such as goals setting, results measuring, and progress reporting. Also, GPRA-USA is an example of the application of NPM in the government bureaucracy (See: Lynn Jr, 1999; Lynn, 1999).

[4]Presidential Instruction Number 7/1999 regarding the Performance Accountability of Government Agencies *(Instruksi Presiden No 7 tahun 1999 tentang akuntabilitas kinerja instansi pemerintah).*

[5]In Bahasa Indonesia: *Peraturan Pemerintah Republik Indonesia Nomor 8 Tahun 2008 tentang Tahapan, Tata Cara Penyusunan, Pengendalian dan Evaluasi Pelaksanaan Rencana Pembangunan Daerah.*

[6]In Bahasa Indonesia: Musrenbang: *Musyawarah Perencanaan Pembangunan.*

[7]Stakeholder is those who directly or indirectly got a benefit or an impact from the planning and implementation of regional development. This explanation refers to the Government Regulation of the Republic of Indonesia Number 8/2008.

[8]In Bahasa Indonesia RPJMD: *Rencana Pembangunan Jangka Menengah Daerah* is a regional development planning document for a period of 5 (five) years (a short term period).

[9]The Regulation of The Minister of Home Affairs Number 54/2010—About Implementation The Government Regulation of The Republic of Indonesia Number 8/2008—About Phase, Procedures

encouragement to measure and account for community protests in the form of demonstrations. As follows:

> Focus on Investment Climate. Analysis of the investment climate performance is carried out on indicators: crime rates, number of demonstrations, length of licensing process, number and types of taxes and levies, number of local regulations supporting business climate...(The Regulation of The Minister of Home Affairs Number 54/2010: 464).

Unfortunately, when the triangulation was carried out, the facts turned out differently on the community, especially those affected by socio-environmental problems due to development plans and government policies, in this case Galesong community who affected by Makassar beach reclamation and sand mining project. They felt that they were not involved in development planning in their area although they did many protests in the form of public demonstrations (Kompas.com, 2017; Mongabay, 2017a; Tribunnews, 2018; walhisulsel, 2019). As they confess:

> There was never such a thing [community involvement]. However, [maybe] the village head was always involved, and from the village head the information was interrupted, because the village head knew that if it was delivered to the people, the people would definitely reject the program and it will not operate anymore. It is because people automatically know if there is a [sand] mining. The fish will run away [from Galesong's sea] and nothing else can be caught. [Galesong's] people were never involved to talk about it (FN1).

The similar confession is expressed by WALHI-SULSEL, as an NGO that actively advocates for socio-environmental problems, concerning the problems faced by Galesong people and other communities in South Sulawesi communities that are affected by other social-environmental problems. WALHI-SULSEL feels that the tendency of the regional government is only to make artificial participation. What essential is the fact that the formal *Musrenbang* has been carried out, although in the end, views and suggestions from community and civil society organizations are not given a place in government development documents such as in the RPJMD.

> For example, [public hearing] of local regulations, most public hearings are initiated by the government... However, when the government sent invitations to us as CSO (Civil Society Organization), the drafts of regional regulation were mostly unattached to the official invitations. Even if they (government) attached [the drafts], several essential attachments were never included. For example, in the case of a regional regulation draft of spatial plan or a coastal zoning plan, [related to the reclamation case and sea sand mining in Galesong.] They (government) have never attached the coastal zoning map that will be regulated in the regulation [draft]... I mean, they made a draft regulation, and then we [CSO] were invited to criticize and give some input on the draft. Yet, they never gave the next draft, how was the change in the draft that occurred after the public hearing, and the most fatal, it was before they (government) officially ratified [a regulation]. They (government) also never sent the [final] draft document to the public. It took a long time and process to be able to get the final draft to be ratified as a local regulation ... [From this case, it can be seen] involving the community only to fulfill the formal obligations, as though that mechanisms

---

for Preparation, Control, And Evaluation of The Regional Development Plan Implementation. (In Bahasa Indonesia: *Peraturan Menteri Dalam Negeri Nomor 54 Tahun 2010 Tentang Pelaksanaan Peraturan Pemerintah Nomor 8 Tahun 2008 Tentang Tahapan, Tatacara Penyusunan, Pengendalian, Dan Evaluasi Pelaksanaan Rencana Pembangunan Daerah.*

> and models of natural resource management systems had changed. However, it is not. People were involved, but their input was not accommodated, what I said as pseudo-participation … as a result, none of the points from us (CSOs) were accommodated in the policy draft [of government]. [For the government,] the important thing was that there was Walhi's signature in the attendance list [as a formal proof of attendance] (WSS1).

Based on the mismatch finding between regulations and practices, I then guess by using the logic of government's regulations-policy-practice unity. There is a possibility of policy and leadership power playing an important role to make this discrepancy occurred. Moreover, seeing regulations provide a space for local governments to adjust the needs of "government" and "regional" in preparing the RPJMD, this is in line with the view of scholars who assess the implementation of NPM in the field of strategic planning that it tends to be influenced by political power and management interests. It is because, at the end of the strategic planning itself, indicators in strategic planning documents will be the indicators of the government's performance evaluation in front of the public (Aspinall, 2013; Diefenbach, 2005; Saunders, 2006).

This claim becomes stronger when we refer to the regulation of the Minister of Home Affairs number 54/2010—about implementation of the Government Regulation of the Republic of Indonesia Number 8/2008—about phase, procedures for preparation, control, and evaluation of the regional development plan implementation. It shows that the government's performance indicator in RPJMD is based on the government priority programs that are aligned with the vision and mission of the elected Governor (the regulation of Minister of Home Affairs number 54/2010, p: 81–82). In this circumstance, there is an opportunity for the influence of political power or political lobby in policy making to determine government's performance indicators in the RPJMD. It is because this indicator directly measures the success of a governor's leadership period.

5.1.1.2 In Terms of Policy and Praxis

After discussing the criticism of NPM style in local government accountability system and adjustments made by the government to the system to fit with the Indonesian values, there are interesting findings to be studied further. It is related to the influence of leadership power, which is thought to be likely to have a significant influence on the practical execution of regional planning and accountability regulations. This finding is based on the fact that the current regional government accountability system is based on self-drafting, self-organizing, and self-reporting patterns with the support of the current Indonesian government model, which gives more room for autonomy to local governments (Djamhuri, 2009; Harun et al., 2013; Sanusi, Butler, & Rola-rubzen, 2013). It is true, as discussed earlier, that bottom-up participatory space is also presented in regulations that bind the local government accountability system. However, the achievement of government performance reported is believed as a standard of the government's success in society, and the Indonesian political system provides a space for regional leaders (Governors, Regents, and Mayors) to be reelected for the second period. Then, the space for an "adjustment" of the achievement standard of government performance due to political and interest orientation is wide open (Aspinall, 2013).

Furthermore, if I relates it to the application of the tiered performance assessment, it determines the career of employees in NPM system (Diefenbach, 2009), which is also applied in the local government accountability system in Indonesia. The opportunities for the presence of performance indicators "adjustment," which consider the ability of personnel, from leaders, middle managers, to employees, can occur. As stated by the following research informant:

> Performance agreements are no longer in echelon 2. In the last two years, it must be individualized, like the integrity pact of a performance agreement between my direct supervisor and me, the staff, and the head of the subdivision that oversees the staff's works. It should have been there, as, in these two years, the KEMENPAN-RB has asked for. That is our standard for work. Thus, in the final results, all [staff] should prepare the reports. In SKP (government works unit) right now, the integrity pact of a performance agreement is a must that based on what they [staff] are working on and what is promised. The staff start to compile the performance report, yearly and monthly...Therefore, staff work according to the agreement in the beginning...It is clear what the staff should do. It is because all tasks have performance measurements (GOS1).

> We also got data [for LAKIP compilation] that already processed [by related agencies] ... In my opinion, indeed, [the information related to development effects] it should be written [into LAKIP]. However, we also got data that already processed [by related agencies]. I argued that please [the related agencies] report all things, whatever it takes. If they face some obstacles, I will fix the words. Thus, I would not let the words make the government's image down. They [related agencies] think the essential things are achieving their targets. Even if not achieved, but at least close to [achieve the targets] (GOS1). (Related to the adjustment and preparation of the accountability report).

Moreover, if the preparation and reporting of government performance accountability are related to the budget and regional economic growth, efficient-effective logic, and regional stability for economic growth, these standards will undoubtedly be the primary consideration in the preparation of performance and accountability indicators. It is as shown in the following guidelines for the preparation of the RPJMD, which links the government's leading programs and budget (Table 5.1).

Furthermore, related to the stability and security of the regional economic and business development, it can be seen in The Regulation of The Minister of Home Affairs Number 54/2010: 464). As follows:

> **Focus on Investment Climate**. Analysis of the investment climate performance is carried out in indicators: crime rates, number of demonstrations, length of licensing process, number and types of taxes and levies, number of local regulations supporting business climate...(The Regulation of The Minister of Home Affairs Number 54/2010: 464).

Interestingly, from that regulation, it can be seen that demonstrations, as one of the common things in a democratic country and guaranteed by law as a sphere of delivering public aspirations, are placed within the scope of an investment climate assessment together with other business indicators. In my opinion, this again illustrates that the orientation of economic development has been robust since the logic of regulations and guidelines for the government's accountability system.

Table 5.1 The correlation between the government's main programs and budget (from the Regulation of the Minister of Home Affairs Number 54/2010: 332)

Code (*Kode*)	Government's affairs and development priorities programs (*Bidang Urusan Pemerintahan dan Program Prioritas Pembangunan*)	Program performance Indicators *Indikator Kinerja Program* **(outcome)**	Performance Conditions at the Beginning of RPJMD (Year-0) *Kondisi Kinerja pada Awal RPJMD (Tahun 0)*	Program's Performance Standard and Budgeting Framework (*Capaian Kinerja Program dan Kerangka Pendanaan*)												The responsible Agencies (*Dinas yang bertanggungjawab*)
				Year (*tahun*)-1		Year-2		Year-3		Year-4		Year-5		Performance Conditions at the end of the RPJMD period (*Kondisi Kinerja pada akhir periode RPJMD*)		
				Target	Rp	Target	Rp	Target	Rp	Target	Rp	Target	Rp	Target	Rp	
1	Mandatory affairs (*Urusan Wajib*)															
1.01	Education (*Pendidikan*)															
1.01.01																
……																
2	Non-mandatory affairs (*Program Pilihan*)															
2.01	**Etc**……															

The orientation of government policies that are more inclined towards the investment-based economic development is also felt strongly by WALHI-SULSEL in their advocacy work:

> Yes, indeed, they [government] are more welcome to investment, rather than the public. For example, the most real one is the matter of mining permit areas (WIUP). Until now, and I believe, not all the people in South Sulawesi have ever read, ever knowing where the mining business permit area that was initiated by the government...Indeed the WIUP is intended for the large-scale business... so they [businessmen] can determine which areas as their target areas for the business activity plans ... Secondly, I still believe that revisions to the spatial planning in [South] sulawesi government, was very easy to do when it was related to a proposal from businessmen rather than from the community. For example, if we talk about spatial planning, we will always divide by two — one as the main area, one as the cultivation area —. I guess, when there is a desire of entrepreneurs to invest in the main area. The government is so quick to change the main area into a cultivation area. Well, that means, if I look at our government's paradigms, the government still adheres to the development paradigms rather than equality paradigms. When it comes to ideological issues, I think the pro-investment paradigm is still strongly embedded in our government. Thus, the public information is restricted, then access to investors is strengthened, giving complete information to businesspeople is more prioritized (WSS1).

In fact, if it refers to the annual report of the advocacy work result of leading socio-environmental NGOs as discussed in the previous chapter, all these NGOs agreed that the government's strong orientation towards economic growth, specifically by encouraging massive investments in the natural resource sector and supporting land clearing for infrastructure projects, is the root of social-environmental conflicts and environmental damage faced by Indonesian society in the last decade (KPA, 2016, 2017a, 2018, 2019; LBH-Jakarta, 2015, 2017a, b, 2018; WALHI, 2018, 2019, 2015, 2016).

Thus, it is not surprising if the general socio-environmental issues that become a public concern has not got a proper place to be discussed in the accountability report of local governments. It can be seen from facts found in other governments' reports (see again Table 4.3 in Sect. 4.2), which face widespread protests from the people over their development policies that tend to harm the people's environment.

Finally, from all the discussion above, several things can be concluded. First, the implementation of local government accountability systems in Indonesia cannot be separated from the development of neoliberalism and its supporting governance, NPM, which is campaigned to the Indonesian government through an economic improvement agreement with the IMF (Harun et al., 2013). The implementation of NPM styles in governance in Indonesia was also supported by the momentum of the public's desire to carry out bureaucratic reform after the 1998 crisis (Zuhro, 2016), and at that time the style of NPM became famous around the world (Diefenbach, 2009; Lee & Haque, 2006; Torres, 2004).

Secondly, in line with Tan Malaka's value, which encourages adjustments to the foreign governance management model adopted in Indonesia, the Government of Indonesia, in real terms, has made adjustments to the SAKIP model, which was originally adopted from USA, adding Indonesian's values. It can be found officially written in the supporting regulations. However, in practice, the implementation of regulations tends to focus on economic development orientation, which gets the main portion in the praxis of SAKIP implementation to the final report, LAKIP. Meanwhile, the social-environmental issues, which are faced by the community as a result of government policies that encourage regional economic development based on leading sectors of natural resources, tend to [even] not given a place as things that should be accounted for in the LAKIP.

Third, the condition then presents a strong suspicion of the strong influence of political interests in determining performance indicators of local governments. This situation can be understood as a consequence of the application of NPM values in determining the government's strategic objectives to practices and accountability report. As the application of NPM presents a self-governance regime, the government sets its performance indicators by itself. Although, it also provides space for public participation [which tends to be artificial and limited], which, in the end, will be an indicator of a government regime' success. In other words, the opportunity for the presence of performance indicators slack, namely the government tends to compile performance indicators under their interests, is wide open.

5.1.2 *Promoting the Values and Voices of Communities that Are not Accommodated in the Government Accountability System*

Furthermore, after criticizing the pattern of NPM in the government performance accountability system, the next important thing about the value of Tan Malaka is to promote the Values and Voices of the Communities that were not accommodated in the Government Accountability System. From the results of this study, I found the fact that government regulations open a wide space for participation from the community to determine the government's development plan. However, in practice, the government is unlike to accommodate and present people voices (one research informant said: pseudo participation), which is generally pitched in protest and rejection of government development programs that tend to bring social-environmental problems. It is then interesting to dig deeper. This subchapter will then focus on discussing this study's findings of examples of community voices that are not accommodated in government accountability systems.

5.1.2.1 Value: Management of Environmental Sustainability with Local Wisdom

WALHI-SULSEL, one of the informants of this research and as an organization that is always at the forefront to advocate social-environmental issues in South Sulawesi, has suggestions regarding how the government should manage and account for environmental issues. Its suggestions come from information on local wisdom in the areas that become the focus of its advocacy. Interestingly, the local wisdom documented by WALHI, if examined more deeply, brings the sustainable and equitable spirit of environmental management.

One interesting example is related to the issue of sand mining in Galesong coast related to reclamation in the city of Makassar. WALHI found the fact that the fishing communities in the Galesong coast have traditionally great attention and local wisdom to the sustainability of their coastal environment. It is because it directly in line to the sustainability of people's livelihoods. The Galesong fishing community shows manifestations of concern for the sustainability of the coastal environment in the form of maps of the fishing area and fishing time[10] (WALHI-SULSEL, 2019:7–10).

Uniquely, at first, the map of fishing area did not use images. Galesong people only hand down maps of the fishing area from generation to generation using verbal information and in everyday fishing practices, and WALHI-Sulsel then helps the community draw the map. As stated by the informant from WALHI-SULSEL:

> For example, in Takalar, the sea of [Galesong] Takalar, 10 miles from the shore. It is a management area. It is a fishing area, which people have their local wisdom way of catching fish there. The community has a specific time to catch fish in the area because they know that fish also need time to breed. There are certain times, where squids will come to gather at that place. There are certain times when they should not go to the sea. It is because dolphins are migrating from Sulawesi to Australia. In these areas, people have their wisdom. People know when they must go to the sea, when they do not go to the sea. We have inventoryed the catchment points of the communities, and we have mapped them, as well as we have overlaid them with maps of mining and coastal zoning plans that are being discussed at the DPRD (province legislative chamber) [together with the government of South Sulawesi Province]. It turns out that the fishing area turned out to be in the mining area. What it means, for thousands of years people have adopted a fishing model that is wise and following their culture and customs. However, it was immediately destroyed by the sea sand mining area, which was initiated by the provincial DPRD [and government]. That is why we are angry. Moreover, why fishermen are willing to go down to Makassar for a demonstration. Because the fishing area [as] their identity and local cultural wisdom has been disturbed (WSS1).

Furthermore, from WALHI-SULSEL information, it can be understood that community involvement is indeed needed from the earliest stages of development planning, in this case including spatial and regional planning,[11] which will ultimately influence the drafting of the RPJMD and local government performance indicators. As can be seen in the case of Makassar city coastal reclamation, it has the consequence to the presence of sea sand mining to hoard the reclamation area and it also has a direct impact on the environment and people's livelihood.

---

[10]The map of traditional fishing areas of Galesong's Fishermen that is disturbed by sand mining can be seen through this link: https://www.walhi.or.id/catatan-akhir-tahun-walhi-sulsel-degradasi-lingkungan-pengaplingan-ruang-hidup-rakyat-terus-meningkat-keselamatan-rakyat-terancam in page 7–10 (WALHI-SULSEL, 2019).

[11]The related regulation, with the Makassar City coastal reclamation issue and the sand mining in the Galesong region, Takalar, is the Regional Regulation Plan (Ranperda) concerning Zoning Plans for Coastal Areas and Small Islands (RZWP3K). WALHI and the Galesong coastal community and fishers association (FORMASI-NEGARA) until this book has done still struggling to eliminate the zoning of the Makassar reclamation space and sand mining in their traditional livelihood areas.

Therefore, the substantial community involvement (not just a pseudo-participation) from the initial planning stage, in my opinion, will increase the chance of a win–win solution between the interests of the government and the people. This is also recognized by WALHI-SULSEL from their advocacy experience so far:

> Well, I am also sure that the community is not 100 percent anti-investment. However, they made sure whether the investment interferes with their livelihood area and their life, and the lives of future generations. Well, if the investment interferes, the community will automatically oppose the investment project. Therefore, the most important thing for me is from the planning aspect. In this planning step, it is then known which ones cannot, which technology the project should use, how should the project use the exploitation model. Hence, people's lives are maintained, and investment and regional income are also maintained. (WSS1).

5.1.2.2 Value: Sustainability of Community Livelihoods

From the previous discussion, it is obtained important information about the relationship between the environmental sustainability and sustainability of community livelihoods. By looking at the relevance of these two themes, in my opinion, the next important value to be presented as an indicator of government performance is how a development project impacts on the sustainability of people's livelihoods. This suggestion can be said as the moral responsibility of the government to the community. This suggestion will further strengthen the legitimacy of the government in society (Archel, Husillos, Larrinaga, & Spence, 2009). As stated by Galesong people who have been affected by reclamation and sand mining, although they have been annoyed by the reclamation and sand mining case, they still put hopes in the government, as regulator, and the involved-companies to improve the sustainability of their livelihoods.

> In the past, companies had said that they would fix it and provide compensation. Basically, anything that could restore people's condition. [it is because that] they had already taken the sand. Thus, we sat down and listened. We [initially] also believed in that [promise]. Imagine all the sand here is all in CPI, which covers 157 hectares. Incidentally, at that time, because the environment had already been damaged, we asked [that the companies] build a fish house, and the fish house program was also one of the (Bupati) Regent's programs … a proposal was also made to be agreed by the Regent. After the Regent agreed on the proposal, we submitted the proposal to the companies. Therefore, they can make it. Because in the beginning, the companies had promised to do anything to fix the damage. However, the result was the proposal was rejected by the company even though they had promised. They prefer to give money to the village head. How are we not angry? The sand was gone, and they had promised in the beginning to help repair the damage caused by their act, and it was denied. We were denied on what we wanted. They prefer to give money to the village head, and even the village head had no help at all from the money he got, even though, there was help in the form of streetlights offered, but what is for? What is the connection between the coastal people who cannot go to sea with streetlights? What is the use of streetlights? How people are not furious! (FN1).

The suggestion relates to government performance indicators on how the impact of a development project on the sustainability of community livelihoods. It has also been the focus of work for socio-environmental NGOs in Indonesia, for example, a report on the impact of eviction on citizens is well presented by LBH-Jakarta (2015, 2017a, b, 2018), as explained in the previous chapter. Even interestingly, it is also in accordance with SDGs campaign, which Indonesian government is also involved in ratifying it,[12] which encourages development to be in harmony with sustainable environmental management for the benefit of the entire community. Furthermore, if the government presents performance indicators on the impact of development on a socio-environmental issue that receives public attention in its accountability report, of course, it will be very good to maintain legitimacy (Archel et al., 2009; Lanis & Richardson, 2013) and can be a tool for people to cross-check the development information in their area.

5.1.3 Presenting the Community-Based Social Values in the Government Accountability System

Previous sections have discussed the values advocated by NGOs and communities affected by socio-environmental issues due to government development plans. So, in this section, I will discuss how to present social and environmental standards based on the values and voices of the people in the government accountability system.

The exciting thing then, when referred to government regulations (see again: Sect. 5.1.1), the space to present values or indicators based on regional needs is highly wide open. However, the next challenge is how to get indicators that originate and relate directly to issues faced by the community can be presented. It is because the challenge lies in the competing power of interests [which tend to be political] from various government stakeholders and interests of the government itself, which also tends to maintain the quality of its performance. Thus, the elections of issue and indicators, starting from the issues discussed in the *Musrenbang*, indicators of government performance as outlined in the RPJMD, until they appear in regional accountability reports (LAKIP), are difficult to be free from the possibility of performance-indicators slack presence.

Based on the argument above, in my opinion, if we learn from the history of social change occurred in Indonesia, one of the optimal ways to present the community-based values in the government's accountability system is to continue the people's organizing movement and active campaigns cross-sectoral (Gordon, 1998; Hadiwinata, 2003; Vatikiotis, 2013). The people's organizing movement and campaign, as carried out by elements of the people and NGOs so far, plus the support from the scholars, will bring counter-discourse to the public. It will also slowly change the government's accountability system in a new direction that gives more space to community-based values (indicators). Interestingly, the analysis based on active political movements like these are unlikely to be the focus of study in the field of

[12]For additional information related to SDGs in Indonesia (see: https://www.sdg2030indonesia.org/#modalIconDefinition & https://www.id.undp.org/content/indonesia/en/home/sustainable-development-goals.html).

socio-environmental accounting (Spence, Husillos, & Correa-Ruiz, 2010). Therefore, the discussion of how the community fights for its local wisdom values and the sustainability of their environment will be one of the essential findings of this research. Regarding the pattern of this people movement and its relationship with development of a model of government accountability will be discussed deeply in the next sections.

5.2 Synthesis of Research Results: The Importance of Representation of Every Community's Class and Group to Be Given a Place in the Government Management System

As discussed previously in chapter three, one of derivatives of Tan Malaka's critical methodology is the importance of representation of every community's class and group to be given a place in the building of a government management system. This value comes from Tan Malaka's view of the interrelationship between the representative models of people's voices in the governance of a country. Tan Malaka believes without the presence of people's representatives who deliver the people's voice, then the practice of government will be far from the welfare of the people. As he said in his book:

> If we think again, members of the Council came from a high class that has no interference with the people, feel the trouble of the *Kromo* (little people). There is no doubt to say that members of the Council are not people's representatives and we are not surprised if [their] regulations contrary to the needs of the people (Malaka, 1921:6).

Indonesia, as a country in which Tan Malaka also fought for its independence, from the beginning of its presence had been declared as a democratic country that respected the voices of its people. This is reflected in the basic philosophy of the state, Pancasila, which promotes democracy run by the wisdom in a consensus or representative of Indonesian people and social justice for all people (Latif, 2011), to the regulations, as discussed earlier.

However, as the focus of this research, if it is associated with the real conditions faced by people who affected by development plans, it still leaves the problem that the aspiration of affected people is not being accommodated in policies to government practices. Even interestingly, as explained earlier, it is because the determination of the strategic plan is very thick with the struggle for political interests. NGOs, which are actively advocating for these issues, tend to find the fact that government is more inclined to the interests of the investor class and the reasons for development than the interests and rights of the affected people (KPA, 2016, 2017a, 2018, 2019; LBH-Jakarta, 2015, 2017a, b, 2018; WALHI, 2018, 2019, 2015, 2016). This condition is also confirmed by WALHI-SULSEL, one of the informants of this study:

> if using class representation, the government is more likely to listen, more likely to agree on information or input submitted by investors … when there is a discussion between the

community and investors, the government is more likely to take references, take information, take data offered by investors, rather than the community ... while the community gives input to the government, there are only two orientations. First, their activities are not disturbed. Second, the environment also does not come polluted or damaged. While the miners [investor] when giving input to the government, their orientation varies. Moreover, the one that is always considered important by the government is that their activities will increase regional income and state revenue. Therefore, [in my opinion,] the government tends to use reasons from entrepreneurs rather than the public. It is because their interests met between government and investors. The investors prepare money and promise the increasing of regional income, while society does not (WSS1).

Then, it is also reflected in local governments's accountability instruments, which tend only to show the economic growth information and do not present information related to the effects of development itself as the findings of this study (see again Tables 4.1 and 4.2 in Chap. 4).

Based on the mismatch between reality and practice, moreover, it tends to harm environmental sustainability and the economic resilience in the affected community. Then, there were protests from the affected group of people. Uniquely in Indonesian context, in the protest movement over government policies and practices, these affected communities often in a coalition with socio-environmental NGOs to advocate these problems, as also found in this study.

When referring to the history of Indonesian social movements, the coalition of NGOs and affected communities based on an issue caused by the Indonesian socio-political landscape is unique. During the new order (*orde baru*) from 1966 to 1998, Indonesia was ruled by a military-authoritarian regime which had done a political propaganda to create one ideology in people's daily life aimed to support the new order. In that 32 years span, there was simply no place for any critical or opposition political movement. Thus, people has lost their critical function to the government (Mietzner, 2012; Noor, 2016; R. William Liddle, 2013). In line with the increasing number of the middle class in Indonesia in 1990s, several critical intellectuals united and set up NGOs for fill the gap as a critical opponents to government in a softest way (Gordon, 1998). Then, these NGOs had filled this critical position, by actively accompany, educate and advocate the community to deal with socio-politic-economic problems from local to central region of Indonesia.

As time went by, these NGOs have become the people's mouthpiece and actively involved in overthrowing the new order regime (Gordon, 1998; Hadiwinata, 2003; Vatikiotis, 2013). Interestingly, after 1998′s reformation, NGOs still choose to fill in the position as a critical-partner of government policy (Hadiwinata, 2003), mainly because remnants of the new order are still being felt in the society's life. The community is still a society without ideology known as a floating mass which does not have a clear ideological foundation to face the government's unsatisfactory policy (R. William Liddle, 2013; Slater, 2014).

Today, NGOs fill this space by taking a firm position as a watchdog to protest unfavorable government policies. Their protests are mainly about current Indonesian government's development policies that are focusing more on the infrastructure and extractive industries which are likely to generate triangular conflicts (among the state, companies, and people) and damaging Indonesian ecosystem (KPA, 2017a;

LBH-Jakarta, 2015; WALHI, 2018). In addition to those reasons, coalitions between communities affected by socio-environmental issues and NGOs often occur as these communities are in sudden shock that their living areas become victims of development without prior consultation, and this situation brings confusion as to who will hear their voices. As stated by WALHI-SULSEL:

> Why there is always environmental conflict. Well, it is because of this, since development planning, the community has never been involved by the government. The community has never been informed, told which locations will be the priority of development for the next five years, therefore, when the investment activities come in and do new exploitation, the community knows, they are just shocked, and [together] rejected it (WSS1).

From the discussion about the connection between people's protests and their advocacy activities with NGOs to fight for the sustainability of livelihoods and the environment, in my opinion, it can be said, in the future, that advocacy activities are alternatives to raise the voice of the people. Also, as it is in the advocacy process, interests of affected communities are fought for, too. Hence, I believe. By understand the advocacy process carried out by NGOs together with the community, it will produce and emancipate a new understanding to the voice (values) of the people that should be presented in policies, practices, and accountability practices of government.

5.2.1 Understanding the Advocacy Process: Understanding the Voices of Unrepresented People

In understanding the advocacy movement of NGOs with the community, it is essential to understand what is "advocacy" first. Etymologically, advocacy is rooted in the word "to advocate" which means "to support something publicly." It also means "the giving of public support to an idea" and "a course of action or a belief" (Oxford dictionary, 2010). From these definitions, it can be said that the word "advocacy" is broad, not limited to just defending, but can also means "to promote" and can even be interpreted as "making changes in an organized and systematic way" (Holloway, 1998).

In Indonesian context, the NGO advocacy movement is often seen as a non-political and non-violence movement by utilizing all the representatives of democratic system tools and legal instruments that available to change public policy (Topatimasang, Fakih, & Rahardjo, 2005:26). Thus, the orientation of NGO's advocacy activities is to focus on a central issue (public policy), not to sporadically revolutionize the entire state system (Topatimasang, Fakih, & Rahardjo, 2005:33). NGOs' advocacy movements tend to have the same perception that they believe changes in government and community systems can be achieved step by step through community organizing and interventions to public government policies that have a broad influence on people's lives (Fig. 5.1).

These conditions are then reflected in the activities of NGOs in Indonesia, as this study put as the example (see again: Sect 4.4). There are three NGOs that this study

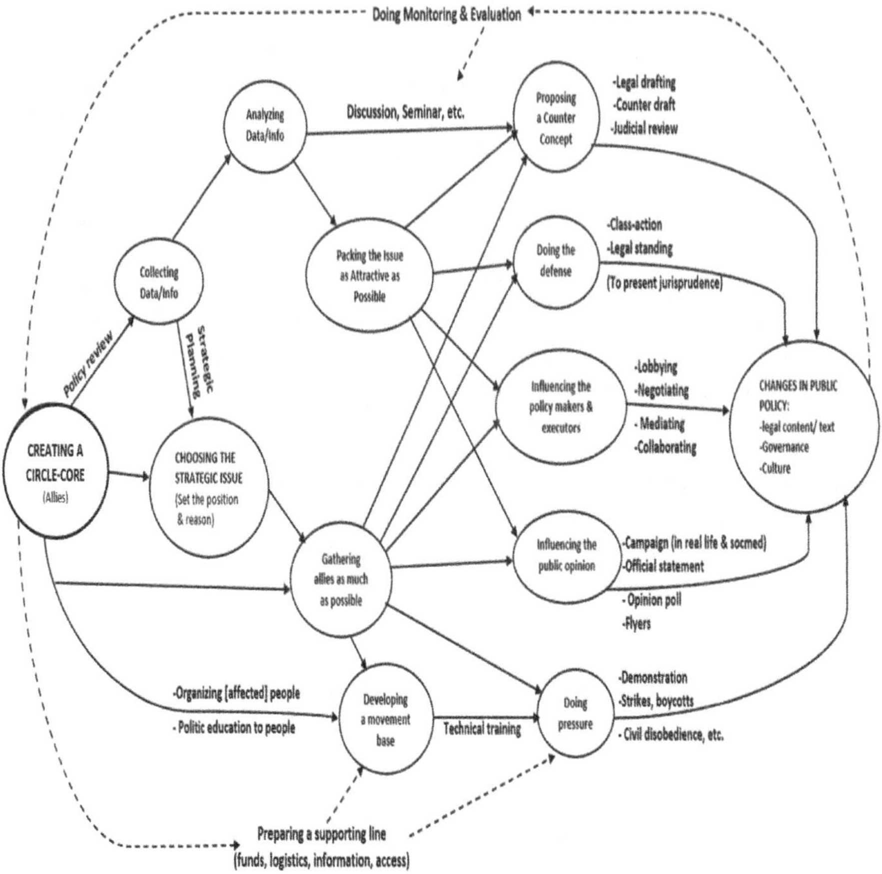

Fig. 5.1 Advocacy Process of NGOs in Indonesia (developed from Topatimasang, Fakih, & Rahardjo (2005:15))

used as examples (WALHI, KPA, and LBH). These NGOs have their uniqueness in each central issue they advocated. WALHI, with its environmental issues (WALHI, 2015, 2016, 2018, 2019), KPA with its agrarian reform issues (KPA, 2016, 2017a, 2018, 2019), and LBH with its human rights issues (LBH-Jakarta, 2015, 2017a, b, 2018). Interestingly, from the different focus of advocacy on the three central issues, these three NGOs have several common perceptions related to what the causes and what can prevent the occurrence of social-environmental issues in the community that they find in their advocacy process.

First, from their advocacy activities on these central issues, they found that the government's development orientation tends to focus on economic growth as well as supporting market (investors) interest (taste of [neo] liberal). It is the root of Indonesian social-environmental problems in the past decade (WALHI, 2015: 3; KPA, 2018:24; LBH-Jakarta, 2015:1). It is because this perspective tends to marginalize interests and local wisdom of other communities, especially the urban poor, farmers/fishermen, and indigenous peoples.

Second, they believe, it is important to involve the people in real terms as equal to the government, not just mere ceremonial socialization, in the planning to the implementation stage of development in the region. It then becomes the key to avoid development conflicts that tend to seize people's living spaces and do not maintain environmental sustainability (WALHI, 2016: 13; KPA, 2018:24).

Third, WALHI, KPA, dan LBH (LBH-Jakarta, 2017b:23), have the same beliefs related to the importance of transparency from policy formulation to government development practices. Even WALHI and KPA stressed their belief that transparency of government development policies and practices is the key to avoid environmental conflicts.

> …Encouraging transparency and the open system of natural resource licenses will be a part of breaking the tangled threads of the ecological conflict. WALHI's advocacy room is massive enough to use court instruments as a tool to demand ecological justice… (WALHI, 2016:13).

> … Open information to the public and evaluate concessions of State and private plantation companies (HGU, HGB, location permits, etc.) Therefore, it could encourage a transparent land system. This openness can also be a way for conflict prevention and resolution and a plan to reduce the inequality that has occurred (KPA, 2019:6).

Finally, WALHI, KPA and LBH encourage law enforcement for the achievement of justice in the community. Especially related to the justice of ecological and people's living space (WALHI, 2016:13; KPA, 2019:6; LBH-Jakarta, 2015: 1–2). This argument can be understood as this study also found. In fact, the legal instruments to technical regulations, from the central government to the local government, are actually highly accommodating aspiration of society's elements who struggle for ecological justice. Unfortunately, it is still far in the praxis. There are a lot of strong political interests in fighting. Therefore, the sense of justice is still far for people to feel in their real lives. This issue also always adorns the annual report on the advocacy work of the three leading NGOs in Indonesia.

These four main issues, from the results of the advocacy of the three NGOs together with the community, are the core of the problem of why the representation of community values so far has not received a proper place in policy to accountability practices of government. In the next chapter, this research will discuss about to what extent and how to synthesize and actualize these four main issues in the local government accountability system, as the focus of this research.

5.3 Synthesis of Research Results: How the Social Structures and Agent (Individual/Society) Can Influence Each Other

Previous sections have discussed criticism of NPM style in Indonesian local government accountability system; the voices (values) of people who are affected by development that are ignored by the government; understanding the advocacy process carried out by socio-environmental NGOs, as a medium to understand the core problem, why the voice of the community that facing socio-environmental problems, are not represented in local government's policies, practices, and accountability. From the discussion above, there is an interesting fact that a gap is seen from the unrepresentation of people's voices affected by socio-environmental issues in the local government accountability system. In fact, as a democratic country, Indonesia's government has been supported by regulatory instruments that open up a space for people to participate in bottom-up development planning.

This finding seems to confirm allegations from Tan Malaka, ± 100 years ago, which highly emphasized the presence of active political movements from the components of society to fight for their voices to be presented in government practices (Malaka, 1921). This perspective is also one of the unique values of Tan Malaka's critical methodology that was developed for this research, which looks at the substance of society, and government ideas are clashing each other to influence the regulations, policies, and practices. This thinking is distinctly different from mainstream western social theory thinking that tends to see the social reality in binary-opposition perspective, in structuralism and non-structuralism approaches (Kamayanti, 2016; Mulawarman, 2010; Powell, 1998; Ritzer & Stepnisky, 2017).

As the final section in this chapter, this section will then continue the synthesis of critical values from Tan Malaka's critical methodology related to how social structures and agents (individual/society) can influence one another. Thus, policies and regulations can be based on societies' interests and be the basis to oversee the government and industry performance for the sake of the people. This value is rooted in Tan Malaka's critical view as follows:

> Let us not be deceived if we hear that in a Parliament there are also the proletarian's deputies. They exist, but they have very little influence over the government of the country (Malaka, 1921:28).
>
> The government is categorized over the "lawmaker" (Parliament) and "the one who runs the law" (minister) …… the minister is full of bureaus [the bodies that establish, propose and enforce the law], then "lawmakers" and the one who run the "laws" actually the ministers too. Moreover, [due to] the interconnection of ministerial bureaus with banks and so on (the capitalist that have interests), then a country that has the Parliament [even based on the people's democratic voice], is willing to charge taxes for the land and sea military actions, robbed the land or the treasures of the people, and willing to allow the war, simply, the government is always enthusiastic to follow the Capitalists' will, which are never satisfied for profit. (Malaka, 1921:30).
>
> The two weapons of the workers (people), who will speak for their intentions in the future, are trade unions and [people] political organizations. The first is obligation to take up economic

> weapons, which can only be sharpened, when it is entered politics that is the politics of the workers (people) themselves (Malaka, 1921:33).

From Tan Malaka's thoughts above, two main things can be synthesized. Firstly, Tan Malaka's ideas stand on a perspective that sees the strong influence of the representation of interests from elements of the community in drafting regulations that guide the government's praxis. As Tan Malaka said, the Rules to the practice of government policy tend to favor the interests of capitalists (or in the case of this research it can be read: the interests of investment and economic development). If the connection of interests between the capital owner and the government is very strong, both in parliament and in the internal government itself (Malaka, 1921:30). Interestingly, as discussed earlier (see again: Sect. 5.1.1), in the Indonesian context, the regulations related to the government's accountability system have normatively been adjusted to the needs of Indonesian people by providing space for community involvement (bottom-up approach) in planning to praxis. However, in the regulation implementation, especially in the government accountability instrument, the "sense" of substantial community involvement seems absent. It can even be said to be inferior to interests of investment and economic development.

Secondly, for making the government maintains and facilitates interests of the small people, then the active political movement of the small people is needed, primarily through political organizations that strive for the people's voice. The organizations that will play the role of check and balance partner. Even, these organizations could replace the government because political organizations also provide their cadres who carry the ideology of the organization (Malaka, 1921:33). If it is contextualized today, Tan Malaka's thoughts are very compatible with the modern political point of view that believes the people organizations, through their active political movements, can play a major role to pressure the government to follow public interests of public at large (Anugrah, 2019). The next question is contextualized by this research, how extent the check and balance movement is driven by the NGOs and people in fighting for government accountability on social-environmental issues, to be more pro-people. These two main essences of Tan Malaka's thoughts will be explored more deeply in the next section.

5.3.1 *Rules and Policies Implementation: The Defeat of the People's Interests*

As discussed previously, normatively, the government regulations related to the accountability system from planning, practice, to accountability reports, have accommodated the people's voice and interests (see again: Sect. 4.1 and Sect. 5.1.1)). However, in the praxis, this study found the opposite. These regulations tend only to be rhetoric, or according to informants of this research, it only to be implemented artificially. In specific, it related to how the representation of voices and interests of small communities affected by social-environmental issues due to government development activities were absent from planning, practice, to government accountability.

Based on findings above, there is a strong suspicion of the strong influence of political interests in the determination of local government performance indicators. It is because the government implements a self-governance regime (as a kind of NPM). The regional government compiles its performance indicators by itself, although it also gives a space to public participation [which tends to be artificial and limited], which in the end, the performance indicators will be an indicator of the success of a government regime. In other words, the opportunity for the presence of performance-indicators slack, such as the government tends to compile performance indicators following (to secure) its interests, is wide open (see again: Sects. 4.1, 4.2 and 5.1).

These conditions above indirectly illustrate the defeat of people's interests in regulations implementation. Furthermore, it can be read that interests of the people, especially the small people who face social environmental issues due to development activities, do not yet have the power to compete with interests of the government and investors (as the main supporting agent of government's development).

Interestingly, if it is examined with the critical value of Tan Malaka's critical methodology related to how social structures and agents (individual/society) can influence one another, in my opinion, those conditions are based on the unbalanced position between the people and the current government in governance. In particular, the community and government did not start from the same position on the information and knowledge related to planning, implementation, and supervision of regional development. Although, one of the keys to the presence of bottom-up community involvement is the presence of active participation from the community at every level of the government development process (Barber, 2009; Fraser, Dougill, Mabee, Reed, & McAlpine, 2006). While what happens in reality, the active participation tends to be practiced only in the form of socialization by the government to the community. In forums that should be active consultation media, the government only socializes its development plans, responds only to the aspirations of the people procedurally (marked by the sufficient presence list of representatives of the elements community). In the end, the initial plan government applied (see again: Sect. 5.1). In fact, in practice, it is often worse, people are suddenly shocked because development has occurred without ever involving them as confronted by the Galesong community as one of this research objects (see again: Sect. 4.1). Thus, in such conditions, protests in the form of demonstrations tend to be the shortcuts way for people to convey their demands.

Furthermore, the balanced position between the community and the government with regard to access to information, if it is examined in more depth as a substantial transparency from government to the public, it is still not achieved. It is also the focus of Indonesian leading NGOs' advocacy related to the importance of transparency in the policies formulation, practices, and supervision of government development, particularly, that focuses on the exploitation of natural resources and requires extensive arable land. Also, it often causes social conflict and environmental damage to the area managed by the people (LBH-Jakarta, 2017b:23; WALHI, 2016:13; KPA, 2019:6).

Then, a question arises, how to bring [more] substantial transparency on the planning, activities, and accountability of government development, which in the end, is expected to be a catalyst for active participation from the community. To answer that question, it is interesting to refer to the opinion of WALHI-SULSEL and one of the government staff who is responsible for the government's accountability report. Both agree on optimizing the existing facilities, in this case, website-based information technology and optimizing active participation instruments that have been run by the government. For example, *Musrenbang* and criticism/suggestion boxes (physical or online) in more substantially, no longer operate them formally. This suggestion is reflected in their interview:

> Yeah, like that at the *Musrenbang*, the government should dig what the community really wants, what is the community's need, do not just talk about funds and budget… More often involve the community, and at least once every three months report what the governor is doing…All government agencies that come in direct contact with the community should have a complaint box because in the LPPD[13] there is a template for the suggestion box [or] complaint box, and it should be answered [followed up] not just accepting the advice … Suggestions or complaints boxes are online and offline …And these things have not been reported yet …this [so far] does not seem to be captured because there is no strategic target [related to the suggestions boxes] (GOS1).

> It is still far away, [in my opinion,] an accountable government is a government that considers all information easily accessible to the public. All information that is worth knowing by the public, it should be known by the public without exception. Thus, there are a supervision and control from community. Therefore, the development is indeed on the track. However, what is happening, the government never shows good faith to become an accountable and transparent institution, although almost all public agencies or government agencies already have websites. However, the information on the website is not very satisfying, unable to show that a website is a government tool in publishing information to the public…Now the era of public information, the era of openness, the government must also think about how to provide the broadest possible information to the public. Public information systems must be opened as wide as possible. Especially for those who have genuinely become public information …open mining area business licenses, mining area permit areas, strategic environmental studies, and annual government reports, environmental services and BAPPEDA reports. Let you, the academics, community, even pedicab drivers can access the information. It is because all of those people have interests [to the development of our region] and It should not be complicated (WSS1).

From the discussion above, then I concluded. Substantial transparency, by utilizing all infrastructure channels and information technology, is the crucial starting point to achieve active political movement in the community to participate in and monitor government performance from upstream to downstream. Without the presence of substantial transparency from the government, the people's control on social-environmental issues will remain unbalanced. Government interests will tend to be more powerful. On the other hand, the community will still find it challenging to fight for their interests as access to information on government plans and policies is

[13] *LPPD: Laporan Penyelenggaraan Pemerintah Daerah* (Local Government Implementation Report).

limited. The people will return as a passive drama audience, shocked by the development that suddenly destroys their living space as they never knew and never invited to participate. Also, they just can cry or shout from time to time to protest. Yet, it is always hard to change what happened.

5.3.2 How is Going the Check-Balance from the People?

The previous sections have discussed the importance for the government of presenting substantial transparency as the main starting point for the active political movement of the community to be involved in the planning, implementation, and accountability of government performance. Next, this section will examine in more depth how the active political movement of society so far, in terms of playing the role of checks and balances against the government, and its consequences if linked to the orientation of presenting the substantial transparency.

As explained in the previous chapters, communities and NGOs have been very active in defending the socio-environmental problems caused by government development projects. Unfortunately, people's active movements tend to be reactive. In other words, elements of society tend to move after a problem. As the Galesong community experienced, as one of the subjects of this study, it moves, along with other elements of the community, after being shocked to see a sand extraction project suddenly present in their environment. The closed space of participation and information linked to the development plans of governments in their environment is the cause of this situation (see again: 4.1). The same pattern is also indicated in other regions. Generally, the advocacy movement of the community in fighting for their living space only began after the presence of exploitation activities in the people's living space. After a case occurs, people report and make an advocacy front with NGOs (KPA, 2017a:5).

By looking at those cases, it can be imagined how hard the struggle of people affected by social-environmental issues to fight for the interests and sustainability of their environment. If reflecting the logic of government regulations-praxis, it is certainly impossible to carry out exploitation activities without a legal basis and clear regulations, such as regional regulations on spatial plans and regional or long-term development plans, and exploitation permits. Furthermore, those legal bases are impossible without a long, gradual bureaucratic process. Based on this situation, if people want to change the process of exploitation in their area, of course, their struggle must go on until a regulation is canceled or changed. Admittedly, this is a long and exhausting struggle. This kind of struggle can be seen from the NGOs annual reports, which are the subjects of this research. Their advocacy work with the community on a central issue appears so long and can take years (KPA, 2019; LBH-Jakarta, 2018; WALHI, 2019).

Based on the arguments above, in my opinion, herein lies the relevance to drive the substantial transparency in planning-policies-practices to the accountability of the government by utilizing existing technology and instruments. This suggestion is needed as an intervention to develop transparency in the governance system from upstream to downstream. At a minimum, this new approach can open up new discussion rooms that are more accountable between the government and affected communities, for what is happening in their area. For example, if substantial transparency is implemented by the government in an area experiencing social-environmental issues, the government can use its official website to release official information that can be accounted for about; the extent of development projects, environmental and economic impacts, and steps the government has taken to overcome the negative impact. From the community side, this information can be a guideline for the people to crosscheck the truth and can be a guideline to demand justice if there are indications that regulations or laws have been violated. Meanwhile, from the government side, it will be able to maintain its legitimacy in the eyes of the affected community through the presence of this transparency model.

Furthermore, this approach is in harmony with the theory of convergence, which believes that the development paradigm to policies in the world today can no longer be seen as black and white. Convergence among various approaches is needed to produce a synthesis that is beneficial to humanity (Evans & Stephens, 1988). As this study found, it is essential to develop a system of accountability from government activities, which is integrated from planning, praxis, to accountability report with a spirit of substantial transparency and utilizing available facilities and technology.

Finally, the development of a framework is necessary to apply the results of this study. The discussion on how to develop this study result in a framework will be discussed in the next chapter.

References

Anugrah, I. (2019). Out of sight, out of mind? Political accountability and Indonesia's new capital plan. Retrieved July 23, 2019, from New Mandala website: https://www.newmandala.org/out-of-sight-out-of-mind-political-accountability-and-indonesias-new-capital-plan/

Archel, P., Husillos, J., Larrinaga, C., & Spence, C. (2009). Social disclosure, legitimacy theory and the role of the state. *Accounting, Auditing & Accountability Journal, 22*(8), 1284–1307. https://doi.org/10.1108/09513570910999319.

Aspinall, E. (2013). A NATION IN FRAGMENTS: Patronage and Neoliberalism in Contemporary Indonesia. *Critical Asian Studies, 45*(1), 27–54. https://doi.org/10.1080/14672715.2013.758820.

Barber, T. (2009). Participation, citizenship, and well-being: Engaging with young people, making a difference. *Young, 17*(1), 25–40.

Diefenbach, T. (2005). Competing Strategic Perspectives and Sense-making of Senior Managers in Academia. *The International Journal of Knowledge, Culture, and Change Management: Annual Review, 5*(6), 126–137. https://doi.org/https://doi.org/10.18848/1447-9524/cgp/v05i06/49910

Diefenbach, T. (2009). New public management in public sector organizations: The dark sides of managerialistic "Enlightenment." *Public Administration, 87*(4), 892–909. https://doi.org/10.1111/j.1467-9299.2009.01766.x.

Djamhuri, A. (2009). *A case study of governmental accounting and budgeting Reform at local authority in indonesia: an institutionalist perspective.* USM.

Evans, P., & Stephens, J. D. (1988). Studying development since the sixties. *Theory and Society, 17*(5), 713–745.

Fraser, E. D. G., Dougill, A. J., Mabee, W. E., Reed, M., & McAlpine, P. (2006). Bottom up and top down: Analysis of participatory processes for sustainability indicator identification as a pathway to community empowerment and sustainable environmental management. *Journal of Environmental Management, 78*(2), 114–127.

Gordon, J. (1998). NGOs, the environment, and political pluralism in New Order Indonesia. *Journal of the Southeast Asian Studies Student Association, 2*(2), 1–26.

Hadiwinata, B. S. (2003). *The politics of NGOs in Indonesia: Developing democracy and managing a movement*. Routledge.

Harun, H., An, Y., & Kahar, A. (2013). Implementation and challenges of introducing NPM and accrual accounting in Indonesian local government. *Public Money and Management, 33*(5), 383–388. https://doi.org/10.1080/09540962.2013.817131.

Holloway, R. (1998). Establishing and running an advocacy NGO. *AKDN Civil Society Programme*.

Kamayanti, A. (2016). *Metodologi Penelitian Kualitatif Akuntansi: Pengantar Religiositas Keilmuan*. Malang: Yayasan Rumah Peneleh.

Kirkpatrick, I., Ackroyd, S., & Walker, R. (2005). *The new managerialism and public service professions*. In Hampshire: Palgrave Macmillan. Springer.

Kompas.com. (2017). Tolak Pengerukan Pasir Putih untuk Reklamasi, Warga Takalar Demo ke Makassar. Retrieved August 1, 2019, from https://regional.kompas.com/read/2017/06/11/173 13341/tolak.pengerukan.pasir.putih.untuk.reklamasi.warga.takalar.demo.ke.makassar.

KPA. (2016). *Catatan Akhir Tahun 2016: Liberalisasi agraria diperhebat, reforma agraria dibelokkan*. https://doi.org/10.1017/CBO9781107415324.004

KPA. (2017a). *Catatan Akhir Tahun 2017 Reforma Agraria Di Bawah Bayangan Investasi*. Retrieved from https://kpa.or.id/publikasi/baca/laporan/25/Catahu_2017:__Reforma_Agraria_Dibawah_B ayang_Investasi__Gaung_Besar_Dipinggiran_Jalan/

KPA. (2018). *Catatan Akhir Tahun 2018: Masa Depan Reforma Agraria Melampaui Tahun Politik*. Retrieved from https://kpa.or.id/publikasi/baca/laporan/30/Catahu_2018:_Masa_Depan_Refo rma_Agraria_Melampaui_Tahun_Politik/?lang=id

KPA. (2019). *Laporan Ringkas: Empat Tahun Implementasi Reforma Agraria*. Retrieved from https://kpa.or.id/publikasi/baca/laporan/77/Empat_Tahun_Implementasi_Reforma_Agraria/?lan g=id

Lanis, R., & Richardson, G. (2013). Corporate social responsibility and tax aggresiveness: A test of legitimacy theory. *Accounting, Auditing & Accountability Journal, 26*(1), 75–100.

Latif, Y. (2011). *Negara paripurna: historisitas, rasionalitas, dan aktualitas Pancasila*. Gramedia Pustaka Utama.

LBH-Jakarta. (2015). *Atas Nama Pembangunan: Laporan Penggusuran Paksa di Jakarta Tahun 2015* (Vol. 2017). Retrieved from https://www.bantuanhukum.or.id/web/atas-nama-pemban gunan-laporan-penggusuran-paksa-di-jakarta-tahun-201

LBH-Jakarta. (2017a). *Mengais di Pusaran Janji: Laporan Penggusuran paksa di Wilayah DKI Jakarta tahun 2017*. Retrieved from https://www.bantuanhukum.or.id/web/mengais-di-pusaran-janji-laporan-penggusuran-paksa-di-wilayah-dki-jakarta-2017/

LBH-Jakarta. (2017b). *Seperti Puing: Laporan Penggusuran Paksa di Wilayah DKI Jakarta Tahun 2016*. Retrieved from https://www.bantuanhukum.or.id/web/wp-content/uploads/2017/04/ Laporan-Penggusuran_LBHJAKARTA_2016.pdf

LBH-Jakarta. (2018). *Masih Ada: Laporan Penggusuran Paksa di Wilayah DKI Jakarta Januari-September 2018*. Retrieved from https://www.bantuanhukum.or.id/web/masih-ada-laporan-pen ggusuran-paksa-di-wilayah-dki-jakarta-2018/

Lee, E. W. Y., & Haque, M. S. (2006). The new public management reform and governance in Asian NICs: A comparison of Hong Kong and Singapore. *Governance, 19*(4), 605–626. https://doi.org/ 10.1111/j.1468-0491.2006.00330.x.

Lynn, L., Jr. (1999a). Requiring bureaucracies to perform: What have we leamed from the US Government Performance and Results Act (GPRA)? *Politiques Et Management Public, 17*(2), 1–20.

Lynn, L. E. (1999b). Public management in North America. *Public Management: An International Journal of Research and Theory, 1*(2), 301–310. https://doi.org/10.1080/14719039900000008.
Malaka, T. (1921). *Parlemen atau Soviet* (2014, Ed.). Jakarta: Sega Arsy.
Mason, J. (2002). *Qualitative researching*. Sage.
Mietzner, M. (2012). Indonesia's democratic stagnation: Anti-reformist elites and resilient civil society. *Democratization, 19*(2), 209–229. https://doi.org/10.1080/13510347.2011.572620.
Mongabay. (2017a). Aksi Warga Takalar Menolak Tambang Pasir: Jangan Paksa Kami Menjadi Teroris. Retrieved August 1, 2019, from https://www.mongabay.co.id/2017/06/17/aksi-warga-takalar-menolak-tambang-pasir-jangan-paksa-kami-menjadi-teroris/
Mulawarman, A. D. (2010). Integrasi Paradigma Akuntansi: Refleksi atas Pendekatan Sosiologi dalam Ilmu Akuntansi. *Jurnal Akuntansi Multiparadigma, 1*(1), 155–171.
Noor, F. (2016). Perilaku Politik Pragmatis Dalam Kehidupan Politik Kontemporer: Kajian Atas Menyurutnya Peran Ideologi Politik di Era Reformasi. *Masyarakat Indonesia, 40*(1), 57–74.
Piantanida, M., & Garman, N. B. (2009). *The qualitative dissertation: A guide for students and faculty*. Corwin Press.
Powell, J. (1998). *Postmodernism for beginners*. New York, USA: Writers and Readers Publishing Inc.
Ramli, R., & Nuryadin, P. (2007). Ten years after: Impact of monetarist and neoliberal solutions in Indonesia. *Economic and Political Weekly*, 79–88.
Ritzer, G., & Stepnisky, J. (2017). *Modern sociological theory*. SAGE Publications.
Robison, R., & Rosser, A. (1998). Contesting reform: Indonesia's new order and the IMF. *World Development, 26*(8), 1593–1609.
Rosser, A. (2013). *The politics of economic liberalization in Indonesia: state, market and power*. Routledge.
Sanusi, A., Butler, B., & Rola-rubzen, M. F. (2013). Local Politics and Business in Indonesia: A Non-market Strategy Approach. *Provincial governance strengthening programme supported by UNDP Indonesia* (pp. 1–20). Unika Atmajaya, Jakarta: Unika Atmajaya.
Saunders, M. (2006). The madness and malady of managerialism. *Quadrant, 50*(3), 9.
Slater, D. (2014). UNBUILDING BLOCS: Indonesia's accountability deficit in historical perspective. *Critical Asian Studies, 46*(2), 287–315. https://doi.org/10.1080/14672715.2014.898456.
Spence, C., Husillos, J., & Correa-Ruiz, C. (2010). Cargo cult science and the death of politics: A critical review of social and environmental accounting research. *Critical Perspectives on Accounting, 21*(1), 76–89. https://doi.org/10.1016/j.cpa.2008.09.008.
Topatimasang, R., Fakih, M., & Rahardjo, T. (2005). *Mengubah Kebijakan Publik*. Yogyakarta: INSISTPress.
Torres, L. (2004). Trajectories in public administration reforms in European Continental countries. *Australian Journal of Public Administration, 63*(3), 99–112. https://doi.org/10.1111/j.1467-8500.2004.00394.x.
Tribunnews. (2018). Tolak Tambang Pasir, Begini Pernyataan Sikap Masyarakat Galesong Raya. Retrieved August 1, 2019, from https://makassar.tribunnews.com/2018/03/01/tolak-tambang-pasir-begini-pernyataan-sikap-masyarakat-galesong-raya
Vatikiotis, M. R. J. (2013). *Indonesian politics under Suharto: The rise and fall of the New Order*. Routledge.
WALHI. (2015). *Tinjauan Lingkungan Hidup 2015 WALHI*. Retrieved from https://walhi.or.id/wp-content/uploads/2016/12/OutLook2015_W.pdf
WALHI. (2016). *Tinjauan Lingkungan Hidup WALHI 2016: Keharusan Pembenahan Struktural untuk Perbaikan Tata Kelola*. Retrieved from Wahana Lingkungan Hidup Indonesia website: https://www.walhi.or.id/2016/12/22/outlook-2016-tujuan-lingkungan-hidup/.
WALHI. (2018). *Tinjauan Lingkungan Hidup 2018 : Bagaimana Masa Depan Keadilan Ekologis 2018?* Retrieved from https://walhi.or.id/wp-content/uploads/2018/12/Layout_Tinjauan-Lingkungan-2018.pdf
WALHI. (2019). *Tinjauan Lingkungan Hidup 2019: Membawa Beban Masa Kini ke Masa Depan?* Retrieved from https://walhi.or.id/tinjauan-lingkungan-hidup-2019/

Walhisulsel. (2019). Aliansi Selamatkan Pesisir (ASP) Tolak Tambang Pasir Laut, Revisi RZWP3K Sulsel! Retrieved August 1, 2019, from https://walhisulsel.or.id/2533-press-rilis-aliansi-selamatkan-pesisir-asp-tolak-tambang-pasir-laut-revisi-rzwp3k-sul-sel/

WALHI-SULSEL. (2019). *Catatan Akhir Tahun- WALHI Sulawesi Selatan 2018: Degradasi Lingkungan dan Pengaplingan Ruang Hidup Rakyat terus Meningkat, Keselamatan Rakyat Terancam.* Retrieved from https://walhi.or.id/catatan-akhir-tahun-walhi-sulsel-degradasi-lingkungan-pengaplingan-ruang-hidup-rakyat-terus-meningkat-keselamatan-rakyat-terancam

William Liddle, R. (2013). Improving the quality of democracy in Indonesia: Toward a theory of action. *Indonesia,* (96), 59. https://doi.org/10.5728/indonesia.96.0057.

Zuhro, R. S. (2016). Good Governance dan Reformasi Birokrasi di Indonesia. *Jurnal Penelitian Politik, 7*(1), 21.

Chapter 6
Towards Local Government's Integrated Accountability Framework

The earth, water, and natural resources contained therein are controlled by the State and used for the greatest prosperity of the people.
—Article 33, paragraph 3, 1945 Constitution of Republic of Indonesia

Previously, in chapter five I analyzed and interpreted the result of this study by applying the critical methodology of Tan Malaka. From the analysis and interpretation process, it was found five important values related to emancipation values from elements of society that continue to fight for ecological justice in their living space amid the ecological damage caused by government development policies. These values are: society, altogether with NGOs encourage (1) participation and (2) substantial transparency for the community to monitor government development by optimizing government infrastructures from planning to praxis. Also, they (3) encourage sustainable environmental management based on local wisdom of the community, and (4) ensure the community livelihoods sustainability. Finally, they (5) encourage law enforcement to achieve ecological justice in the community.

These value findings will be further elaborated in this chapter to develop a new integrated accountability framework in local governments. I hope. Eventually, the people's voices will be represented in the government's performance accountability system as it will support the government legitimacy in public eyes. Also, it will support the achievement of sustainable development goals, which also ratified by Indonesia.

6.1 Substantial of Participation and Transparency as the Basis

Based on the research and analysis results in chapter four and five, there are interesting facts that show substantial participation and transparency as the main things that must be presented by the government to materialize sustainable development while maintain government legitimacy in public eyes. These two things have so far been inclined to be missed by the government from the planning to accountability of its

H. M. Shahib, *Towards the Local Government's Integrated Accountability Framework*,
https://doi.org/10.1007/978-981-33-6617-6_6

activities, particularly related to development activities with broad impact on people's ecological living space. Although, it is actually set in government regulations and binding the process of development planning, implementation, and accountability (see: The Government Regulation of the Republic of Indonesia 8/2008 and the law of the Republic of Indonesia 14/2008 about Disclosure of Public Information).

This research synthesis then comes to the phrase "substantial participation and transparency" as a basic offering for the development of integrated local government accountability, which further accommodates voices of people who are often times being marginalized from the space of government development policy to government accountability practices. There are three main considerations why I choose this phrase as the basis to develop an integrated accountability framework for local government.

First, this research finds a fact that people affected by social-environmental issues, and NGOs that actively advocate for environmental and social issues in Indonesia, are experiencing the lack of participation and transparency in government development activities. The community is likely to be aware on a development project in their living area when the project is underway. Thus, the unequal and ineffective form of delivering aspirations by the community tends to become spontaneous reaction to deliver community aspirations (e.g. demonstrations).

Second, based on information from my research informants, in this case NGOs, several facts showing although the government has provided a participation space for community elements to be involved in public consultation rooms or regional development planning meetings, the form of participation still tends to be mere of socialization and formality. That is, the government tends to only describe its plans and gather input from community elements. But in the end, it was still the initial plan of the government that was carried out. The participation of community elements tends to be solely judged based on the attendance list of representatives of community elements (NGOs) in the public consultation room. In fact, in many public consultation forums attended by my NGOs informants, it is difficut to obtain a copy of the notes or agreement between the government and the community elements in a forum. Thus, it is difficult for elements of the community to supervise matters that have been discussed and agreed between the government and the community elements. Then, there is a possibility of the presence of stronger political interests in the regional development planning process due to this condition.

Third, the balance of information is sinking in the public sphere due to these two things above. While in a democratic life, the imbalance of information due to the lack of transparency of government is one of the things that can harm democracy itself (Esser, 2013; Misztal, 2001). It can be said that the lack of transparency of government will cause difficulties for society and the government to find consensus in public space. It is because that is difficult for people to participate due to the uneven information available to them. Even if they are not controlled, the communication obstruction resulting from the lack of transparency and participation will present counter-consensus movements such as demonstrations and boycotts. That is, both will certainly be prejudicial to many parties, and the development process in the middle of the community itself.

From these three considerations, it can be concluded that the need for "substantial participation and transparency" is a must for the government. First, it will maintain a healthy democratic climate. Second, it is also important to support the presence of inclusive and sustainable development in the community.

I chose to continue to compare "participation and transparency" by the analogy of friendship. In a friendship between two people, A and B, none of them takes the lead to decide that the two of them are friends. The same analogy can be applied to participation and transparency in the context of democratic governance, especially in the case of this study. Participation and transparency go hand in hand. Public participation is not possible without the presence of government transparency. Vice versa, it is difficult for the government to be transparent if there is no active [political] participation from the community to encourage the government to be transparent. Whereas, the selection of the word "substantial" that follows "participation and transparency", is a form of criticism from this research for substantive community participation and transparency from the process of planning to the government accountability of its development activities; although, it is already set up by the government regulations.

6.2 Starting from the Planning Stage

The previous section discussed substantial participation and transparency as the basis to present an integrated accountability system from the government's development policy to the government's accountability practices. Indeed, the discussion will continue with how to internalize the participation and substantial transparency spirits in order to develop the current government's accountability system into a more emancipatory towards the voice of the people who are likely to be victims of government development policies.

Referring to the logic of government regulations, it is known that there are links among regulations, planning, implementation and reporting of government activities as a form of government accountability. Interestingly, in Indonesian context, this study found in the post-New Order era, government regulations are actually provided a space for transparency and bottom-up participation from development planning to accountability for government performance.[1] However, in the practice, it still leaves problems, particularly related to development issues with major social-environmental effects. Affected communities and NGOs, which are actively advocating these issues, perceive the new government participation scheme in the post New-order tends to focus in achieving the formal legal obligations. On the contrary, it has not fulfilled the substance of participation itself, which has the consequence of the unrepresented people's voice from the government development planning space to the practice of government accountability.

[1]Refer to the Government Regulation of the Republic of Indonesia 8/2008 and the law of the Republic of Indonesia 14/2008 about Disclosure of Public Information and Sects. 4.1 and 5.1.

Furthermore, this study results indicate the occurrence of an alleged political influence of a government regime that caused asynchronous conditions of government regulations to their implementation. This happens as the current local government lives in the era of self-governance (self-drafting, self-organizing, and self-reporting) with the support of the Indonesian political-government model that gives more autonomy to the regional government (Harun, An, & Kahar, 2013; Sanusi, Butler, & Rola-rubzen, 2013), and provides space for regional leaders (governors, regents, mayors) to be re-elected in the second period (Aspinall, 2013). Thus, the reported achievement of government performance in the midst of society will greatly influence public political perceptions.

Moreover, if we refer to reports from NGOs that actively advocate for social-environmental issues, they find close relations between government development plans and investment interests (WALHI, 2015:3; KPA, 2018:24; LBH-Jakarta, 2015:1). Consequently, the space to "adjust" from planning to government performance indicators, due to the orientation and strength of the political interests of the government elite and investors, is increasingly open.

Thus, this research offers a solution to these conditions. It suggests that substantial participation and transparency should start from the government development planning process. It is because this phase will greatly determine the government's policies, activities, and accountability. Also, it is set on the bottom-up participation of the various elements of society in the *Musrenbang*[2] official forum.

Substantial participation and transparency here are intended to ensure the presence of community elements' active participation and the way to oversee the people's voice so as not to lose from the stronger political interests. This solution was born from this study's dialectical process by the utilization of Tan Malaka's critical perspective as the research analysis tool; particularly, about the importance of presenting a coherence between non-structural aspects (community aspirations) and structural (government governance systems).

Another interesting thing, Tan Malaka's critical perspective highly encourages for adjustments to a system, knowledge, and technology for the sake of people's interests. This can be found in a variety of Tan Malaka's literature, including:

> Throughout our knowledge, it is enough to show the nature, origin, and institutions of the West, one of which we must choose if we have the freedom to govern our own country (Malaka, 1921:42)……All domestic institutions without nationality would undoubtedly corrupt the nation (Malaka, 1921:43).
>
> Although Indonesia is the richest country in the world, as long as science is not independent, like the politics of the country, Indonesia's wealth will not make the Indonesian people happy, even it will only make it difficult … That is how political economy and science are a combination that must not be broken (Malaka, 1943:45).
>
> Admit it sincerely, that you are able and must learn from Westerners. But you should not be a copycat to a Westerner, be an intelligent student from the East who follows the will of

[2] *Musrenbang (Bahasa Indonesia: Musyawarah Perencanaan Pembangunan)* or Development Planning Forum in English, is an inter-stakeholders forum in the context of regional development plans preparation before the RPJMD ratified as the local government's regulation, and become the standards for governments' accountability performance indicators.

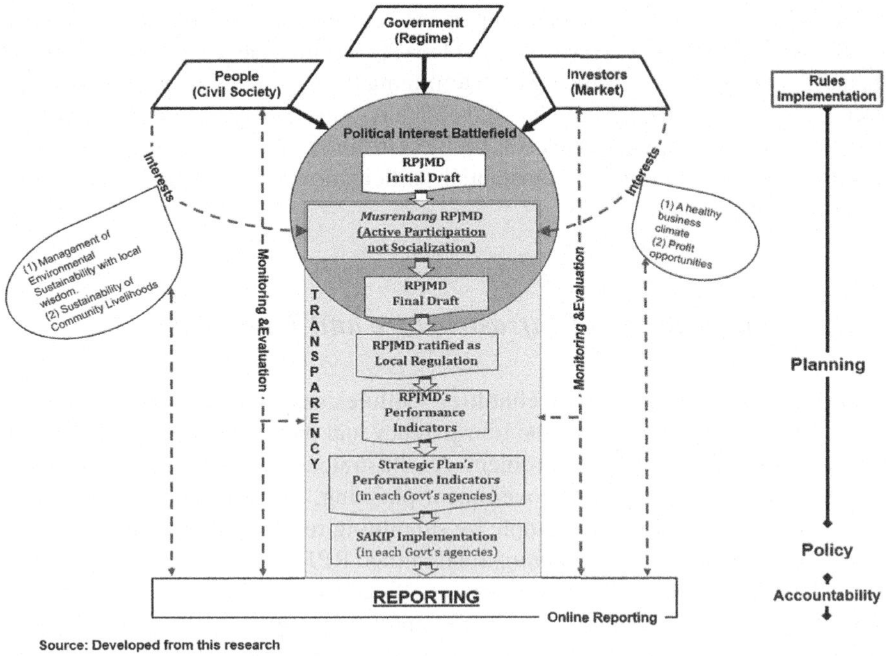

Fig. 6.1 Planning Stage as a Crucial Stage in Government's Accountability System

> nature and so forth that can exceed the intelligence of his/her teachers in the West… Then the skill and willingness to follow the nature may grow, and with it the immeasurable wealth of Indonesian land can be cultivated and used for the nobleness of the Indonesian people (Malaka, 1926:117–118).

When it is elaborated further, the thought of Tan Malaka have similarities with social change theories that were born later, such as the convergence theory (Evans & Stephens, 1988) and the third path theory (Giddens, 2013), which also encourage for adjustments to the political-economic system by taking good things and leave bad things from various theories and practices of social change for a fairer state and world order and bring benefits to the wider community.

Based on the explanation above, this research then tries to redevelop a form of participation and substantial transparency that is integrated starting from the government development planning phase to the practice of government accountability by applicable regulations.[3] The form can be seen in Fig. 6.1 below.

[3]The Government Regulation of the Republic of Indonesia Number 8/2008 about Steps, Preparation Procedures, Control and Evaluation of Implementation of Regional Development Plan, and The President Regulation Number 29/2014 as a legal basis of the system of performance accountability in government agencies *(SAKIP)*.

On Fig. 6.1 can be seen the planning phase, especially the determination phase of the *RPJMD*,[4] is the most decisive phase of local government development activities in the next five years. Also, the phase of determining the *RPJMD* is an arena of political interest among various government stakeholders, especially among interests of the investors and public, as important stakeholders in the local government development. On this basis, the presence of substantial participation and transparency is highly important in this phase.

6.2.1 Optimalization of Infrastructure and Technology

Optimalization of information technology facilities usage is a way that should be done by the government. Thus, the transparency and participation are truly present and touch the substance of the problem. As illustrated in Fig. 6.1, public access to information from all stages of government planning, activities and accountability must be accessed online, for example by submitting results and providing columns for recommendations to *Musrenbang,* final ratified RPJMD, and government performance indicators through local government official website. This will automatically present a pattern of active participation from community elements in the form of monitoring and evaluation (Bonsón, Torres, Royo, & Flores, 2012; Pina, Torres, & Royo, 2010). It will also indirectly let the government gain the sense of being supervised directly by its people (Bonsón et al., 2012; Pina, 2005; Pina et al., 2010). Therefore, it will encourage the government to pay attention and implement the people's aspirations into the government program. As a bonus, the legitimacy of the government will be stronger in the public eyes (Bonsón et al., 2012; Pina, 2005; Pina et al., 2010; Shahib & Risky, 2017).

However, there are important things need to be considered on the online-based government transparency implementation. As what people in general may have known, there are several confidential government informations related to the State's sovereignty and other private informations that are protected by laws. In that regards, the online transparency schemes application must remain in a corridor that is not related to information that may jeopardize state sovereignty, or in Indonesian context, must be in accordance with the law of the Republic of Indonesia No. 14/2008 about Public Information Disclosure. Thus, it is interesting to consider input from one of the informants from KEMENPAN-RB,[5] as the respective ministry that is responsible for regulation and oversee the implementation of the local government performance accountability system, as follows:

> I think there is also a need for a performance indicator dictionary … if we have a dictionary of performance indicators agreed upon together, we only need to open it that I want to do

---

[4]In Bahasa Indonesia, RPJMD: *Rencana Pembangunan Jangka Menengah Daerah* is a regional development planning document for a period of 5 (five) years (short term).

[5]*KEMENPAN-RB (Kementerian Pendayagunaan Aparatur Negara dan Reformasi Birokrasi) or* (Ministry of Administrative and Bureaucracy Reform).

> such activity and what indicators are suitable, right? This is agreed upon by the regional government [and elements of society], that was what I expected (MA1).

Presenting a dictionary or performance indicators standard that are mutually agreed among governments and society elements seems to be a middle way to avoid the disclosure of confidential information related to state sovereignty and other private information stipulated by the law. In addition, the presence of a mutually agreed upon performance indicator standard will improve the quality of communication between the regional government and its people because both parties have the same perception about the regional government performance indicators.

Finally, encouraging substantial participation and transparency by online schemes can be seen as the antithesis of the problem of the absence of space for participation and transparency in government development planning. Especially today, the government already has enough resources and infrastructure to optimize transparency and online-based participation. With online-based participation and transparency, elements of the community then will have the power of information that is equivalent to the government so that it can be more optimal to oversee development issues related to their living space.

6.3 Completing the Government Accountability Report

Sections 6.1 and 6.2 discussed about how to internalize the spirit of substantive participation and transparency in the integrated government accountability systems from planning to reporting stage. This section will then focus more on discussion about the results of this research related to reporting on government performance accountability, as the final stage of the government accountability system as well as the main window for stakeholders to assess government performance achievements.

This study found interesting facts related to Indonesian local government accountability, specifically related to how the local government accountability reporting instrument does not provide sufficient explanation space for social-environmental issues which are the main concern of the public in the regions. As what has been discussed in Sect. 6.1, the actual Indonesian government regulations have given space to bring up accountability on social-environmental issues, from planning through to reporting. However, as (1) accountability reporting on local government performance is very much based on official government development plans in the form of RPJMD which have rigid standards due to the institutionalization of NPM style. Also, (2) the development planning process (RPJMD) is far from transparent and participatory for all people. In addition, (3) there is a strong allegation about political and investment interests influence from the planning process to regional development programs. It then causes the people aspirations, especially those affected by government development projects, are not being presented from the planning to the government performance accountability report (refer to: Sects. 4.1, 4.2 and 5.1).

After synthesizing this study findings, this research then came to an offer to complete the government's accountability report so that it could truly be a reliable representation of the government's performance accountability report. Particularly, it can strengthen the legitimacy of the government in the midst of people who face social-environmental issues caused by government development projects. In the end, obviously, this will further strengthen the position of Indonesian local government as a regional government that has a policy that favorable towards sustainable development goals.

This study offers three main things to refine the current local government's accountability report. All of them are still based on values of substantial participation and transparency. In addition, this study looks at the style of NPM that consciously or unconsciously penetrated the body of Indonesian government agencies and then made the current governance regime from the lowest to highest levels become highly focused to submit reports based on established standards or references (see again Sect. 4.2.3 NPM assumptions: regarding performance management and measurement systems). Thus, the spirit of optimizing the government's facilities and technology has also becomes an antithesis to avoid government officials trapped in the routine of making reports based on performance achievements that tend to be far from presenting people's true aspirations in their accountability reports (Vicente Pina et al., 2010).

6.3.1 Facilitate Accessibility to Government Environmental and Natural Resource Management Reports

This research found a fact, in legal term, the government already has sufficient instruments to control and explain to the public about development and the impact of development on society. These instruments include: Mining and plantation permits, zoning plans and spatial planning as a development guide, strategic environmental assessment reports,[6] and an environmental impact assessment report on a development project.[7] Unfortunately, in the realm of practice, as found in this study, elements of the community, especially social-environmental NGOs, that actively advocating for social-environmental issues, have difficulty accessing such information. In fact, this information is needed to equalize perceptions and control which areas can and cannot be managed by large-scale industry or development projects that have broad effects on ecological security and the lives of the wider community. Furthermore, this information can also be a legal instrument to defend ecological justice, for example when a company's mining or plantation permit is expired, it is obligatory for the company to stop its exploitative activities.

Furthermore, this study found a fact that social-environmental conflicts taken place in Indonesia are generally caused by the lack of government transparency

[6]In Bahasa Indonesia: *Kajian Lingkungan Hidup Strategis (KLHS).*

[7]In Bahasa Indonesia: *Analisis Mengenai Dampak Lingkungan (AMDAL).*

on the environmental management information. Farmers, fishermen, indigenous people who manage their forests, and poor people are the most vulnerable elements exposed to social-environmental conflicts. Specifically, it is related to the government's nontransparency in zoning and spatial planning as well as mining and plantation management permits. Cases of zoning and spatial planning are found often times, without involving community participation: acquiring traditional livelihood spaces and areas that have been handed by people traditional management-style (Walhisulsel, 2019) and also having formal legal permits (LBH-Jakarta, 2015). That was also found in the case of the reclamation of the coastal city of Makassar, which was the focus of this study, which resulted in damage to the living space of fishermen on Galesong coast (Walhisulsel, 2019; and see again Sect. 4.1).

Moreover, social-environmental NGOs have also observed the lack of government information disclosure, especially related to information on mining and plantation management permits, which is one of the reasons why Indonesian people are difficult to be free from environmental disasters such as floods and forest fires. This is due to the lack of disclosure of this information, which results in difficult enforcement of environmental law due to anonymity regarding who must be responsible for environmental damage (e.g., forest fires and flash floods) (WALHI, 2016:13; KPA, 2019:6). In addition, from the perspective of legitimacy theory, without the presence of clear information from an entity, it is certainly difficult for the community to socially punish a company in the form of giving public pressure to be more responsible for their business processes (Islam & Deegan, 2010; Lanis & Richardson, 2013).

Based on the explanation above, there are plenty of negative consequences arise caused by the lack of government's accountability in terms of environmental management information. On such ground, this research suggests the government to optimize facilities and technology to aid the supervision by elements of society on development activities and industries in their managed area. This is important to bring legitimacy that will push the government and related industries to follow the aspirations of the people.

Furthermore, as criticized by this research, in order to avoid government officials trapped in routine reporting (as accountability reports related to environmental management are basically owned by the government), the online accountability model can be an option to facilitate online access to related information of the local government's environmental management. By utilizing the information technology, it can be easily made into a resume and presented in the local government LAKIP, both in the relevant offices (e.g. the environmental department) and the final LAKIP of the local government. Finally, it will improve the quality and reliability of information presented in LAKIP Local government. More detail is shown in Fig. 6.2.

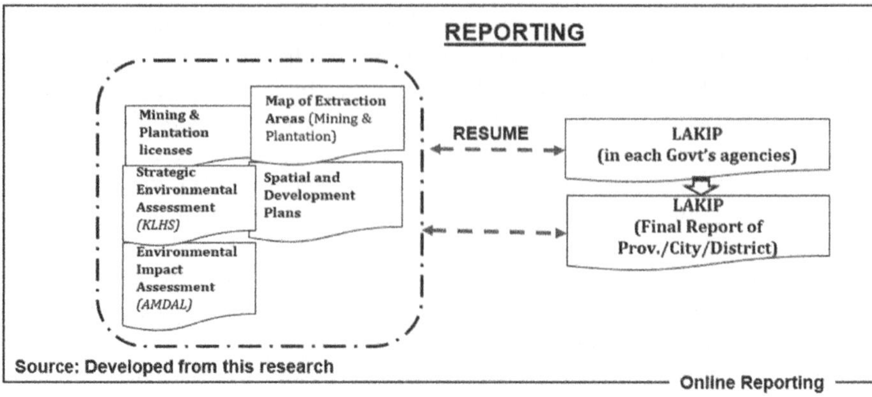

Fig. 6.2 Making easier access to the Government's Environmental and Natural Resource Management Reports

6.3.2 Presenting the Report on the Development Impact of a Local Government Project

Furthermore, it is related to the second offer from this study, which is to present a report on the development impact of a local government project. Surely, it will complement the local government's accountability report. In particular, the presence of this report is an antithesis in facilitating the government accountability report to further anticipate the aspirations of the people, especially those affected by government and investor development projects.

This suggestion departs from one of these study findings which shows the fact that people's livelihood sustainability affected by a development and investment project tends to be missed by the government in the form of a report. This is then essential for the government to convey it. Therefore, the reliable information is presented, where the public can cross-check the truth, and can be held accountable before the law. As this research found those facts, after the implementation of a government development project, affected communities are unlikely to have clarity from the government on the problems that they are experiencing, such as loss of livelihood access, and what steps the government is taking to overcome it.

Even, the solutions offered by the government are often times far from resolving the substance of problems faced by the community. In this case, the sand mining in Galesong coastline, as one of this study samples. The solution to the destruction of fishermen's fishing spaces and lives, presented by the village government officials and mining companies of Galesong sea sand, is to bring streetlights to affected villages (see again: Sect. 5.1.2.2). Likewise, found in NGOs' advocacy reports, the affected communities had difficulty to obtain clear compensation for the running government development project (LBH-Jakarta, 2017:23).

Encouragement to present a report on the development impact is also presented by an informant of this study (from KEMENPAN-RB). In the midst of this regional

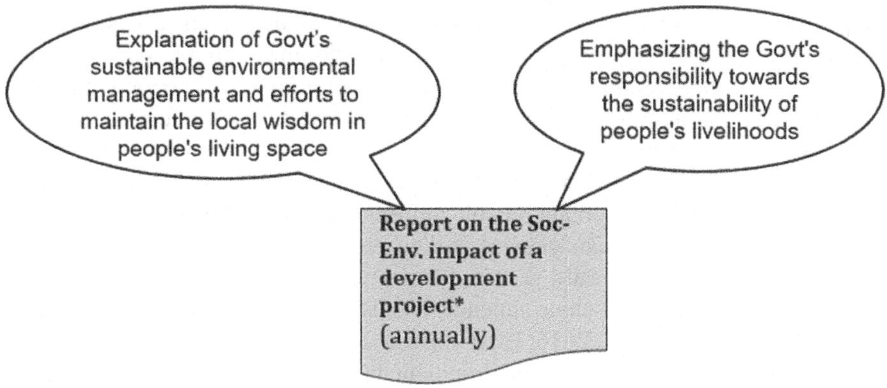

Source: Developed from this research

Fig. 6.3 Framework of Report on the Socio-Environmental Impact of a Government's Development Project

autonomy era, he pushed for doing an in-depth evaluation for every local government development activities with direct impact on society. However, he also admitted that is a rare for local governments to do such practice.

> So, if we look at each of the [development] activities of the government that we want to see the impact on the community. It is necessary to have in-depth evaluation studies for that…However, it is rather difficult because there are more activities that must be done as well as no one has done it. Also, it is rarely [local government] people [want to] do that. Well, it should be BAPPEDA if in local government (MA1).

On such basis, this research came to conclusion about the importance of presenting reports on the development impacts of a local government project. In particular, the report should emphasize on how the government is responsible for sustainable environmental management and its efforts to maintain local wisdom in the community's management space. Also, the report must emphasize the government's accountability towards the sustainability of people's livelihoods. In my opinion, the report should be presented annually during the project development as a medium of official communication for government and affected communities and as a learning tool for future development projects. The proposed framework for the report can be seen in Fig. 6.3.

6.3.3 Presenting the Local Government's Social-Environmental Sustainability Report

This is the final offer of this research, to develop an integrated accountability framework while completing the local government accountability report. This offer has an orientation to present a report on the socio-environmental sustainability of local governments. On one hand, it is expected to play a role as an effort to accomodate the people's aspirations related to their environment management in a sustainable manner. On the other hand, it is expected to provide more reliable information to regional governments in order to manage their environment in a socially equitable manner and to pay attention to environmental sustainability in the regions.

The suggestion to local government to present the socio-environmental sustainability report is the result of this research synthesis. We found the fact that community elements have difficulty to obtain reliable information about what actually happened related to environmental management in their area. Generally, reports on environmental management that easily accessed and often discussed in public spaces are advocacy reports of social-environmental NGOs that are concerned about social-environmental issues (KPA, 2016; LBH-Jakarta, 2015; WALHI, 2018). On the other hand, the public, including social-environmental NGOs, find difficult to have access to information related to government environmental management, specifically on matters related to information on environmental resilience associated with infrastructure development activities and extractive activities of natural resources, such as mining and plantations that have extensive environmental impacts (see again: Sect. 4.1).

This condition then causes difficulties to present information and to reach the same frequency in the discussion about environmental issues within the regions. It is because the government has its own opinions while the community has its different opinions regarding the impact of government development projects on the community's living space. Finally, protests by community elements repeatedly arise in the form of demonstrations or open conflicts with government and project development companies due to ineffective communication caused by the absence of the accepted standard of understanding.

Interestingly, referring to the information from an informant who belongs to WALHI National Executive, the government actually has a legal basis to lead into the presence of regional social-environmental sustainability reports. These include: Government Regulation of the Republic of Indonesia number 46/2016 about Procedures for Organizing Strategic Environmental Studies[8] and Government Regulation of the Republic of Indonesia number 46/2017 about Environmental Economic Instruments.[9]

> These government regulations [number 46 year 2016 and number 46 year 2017] are actually calculate the value of valuation on the environment … but our bureaucracy is difficult. Although the PP [Government Regulation] has already been issued, it will not be able to go down before the Ministrial Regulation, technical regulation, or Technical Guidelines are available… Well, based on the Government Regulation, if it works effectively, more or less each region will have a national name, it has environmental GDP, and in regions it has an environmental GRDP (WNE1).

Taken this fact into account, I then argue that the government could have made an effort to present a regional social-environmental sustainability report using available

[8] *Peraturan Pemerintah Republik Indonesia Nomor 46 Tahun 2016 Tentang Tata Cara Penyelenggaraan Kajian Lingkungan Hidup Strategis.*

[9] *Peraturan Pemerintah Republik Indonesia Nomor 46 Tahun 2017 Tentang Instrumen Ekonomi Lingkungan Hidup.*

information by utilizing its information technology while waiting for the presence of official regulations governing it. That is because in the logic of government bureaucracy, policy intervention can be done as long as it does not violate existing rules. Obviously, this will also cause local governments to consider the costs and benefits. But, with the availability of current information technology, it should not be a problem. Thus, it must present a strong political will from the government and a strong political stimulus from the elements of society to bring this up, one of which is the presence of this research.

This research then comes up with a number of suggestions for local government policy interventions to present social environmental sustainability reports so that they can truly present in the community. This is considerably very urgent to be presented as a medium of government communication amid increasingly widespread environmental damage and protests from the community, as well as causing a decline in people's sense of trust to the government, due to the increasing cases of social-environmental conflict in the community (KPA, 2019).

First, it can be started by optimizing the initial environmental data hold by the local government. This can be done by conducting an initial inventory of what data has been owned by the local government so far as discussed in Sect. 6.3.1 above.

Second, benchmarking and adjusting to generally accepted standards. One of the critical values of Tan Malaka methodology used as guidance in this study is the spirit of adjusting existing knowledge (even it is from foreign knowledge) to the interests of the broadest people. Interestingly, this has also been done by government elements. Take for example, Indonesian Central Statistics Agency (BPS) has adopted the System of Environmental-Economic Accounting (SEEA) framework,[10] trying to integrate the size of economic statistics and environmental statistics (BPS, 2017). To this day, BPS continues to make improvements and continues to open space to receive constructive input on the development of the SEEA in Indonesia (BPS, 2018).[11]

One of the SEEA advantages is that it displays non-financial information, especially macro information related to the resilience of a country's natural resource assets associated with economic activity in the field[12] (BPS, 2019). Unfortunately, this also includes its shortcomings caused by its focus only on information on natural resource assets that are directly related to economic activities. Other information that

---

[10]The SEEA was initially developed by the United Nations' Experts Committee on economic-environmental accounting formed in 2005 and has now begun to be applied in many countries in the world (Stiglitz, Sen, & Fitossi, 2009), including Indonesia.

[11]In Indonesia SEEA is called SISNERLING *(Sistem Terintegrasi Neraca Lingkungan dan Ekonomi Indonesia)* (BPS, 2018).

[12]For detail information related to how the SISNERLING implemented for measuring the Indonesia's natural resource assets associated with economic activity in terms: (1) Indonesia Land Accounts, (2) Timber resources Asset Accounts, (3) Mineral and energy resources asset accounts, (4) Indonesia integrated environmental-economic accounts, just go through this link: https://www.bps.go.id/publication/2019/12/27/53e7360b49c122c58475d2c9/sistem-terintegrasi-neraca-lingkungan-dan-ekonomi-indonesia-2014-2018.html there are two language available (Indonesia and English).

is difficult to measure and related to non-monetary values is not present there, particularly information related to social impacts arising from economic activities in the natural resource or environmental sector (Stiglitz et al., 2009). Of course, this can be understood because the initial orientation of SEEA presence is only as a tool for recording the integration of economic and environmental statistics (United Nations, 2014).

At this point, local government may present its sustainability report based on the initial account held by SEEA or SISNERLING, specifically related to information on the resilience of natural resources in the working area of local governments. Then, the government can add accounts related to social and cultural impact information that is difficult to measure in monetary terms. In my opinion, several reports can become examples for the government to develop measures of social impact caused by development, for instance, the eviction reports from LBH-Jakarta (2015) and agrarian conflict reports from KPA (2016). Surely, it is expected to bring closer or even the same perception among elements of society and government about what we really want to achieve in sustainable development.

Third, to avoid the government is busy in taking care of reporting preparations as a consequence of the application of the NPM model. The use of information technology is certainly needed. One step that can be done is to create one data policy that can be accessed online by report submittee and the public as report users. No more menu choices based solely on the appearance of each related agencies, however it should be based on a summary of the main/current issues as a summary of the current information presented in LAKIP. The three suggestions above are then summarized in the following Fig. 6.4.

Finally, after discussing all the findings of this study about developing a new integrated accountability framework in local governments, then the big picture of the proposed development of an integrated local government accountability system can be illustrated in Fig. 6.5 below:

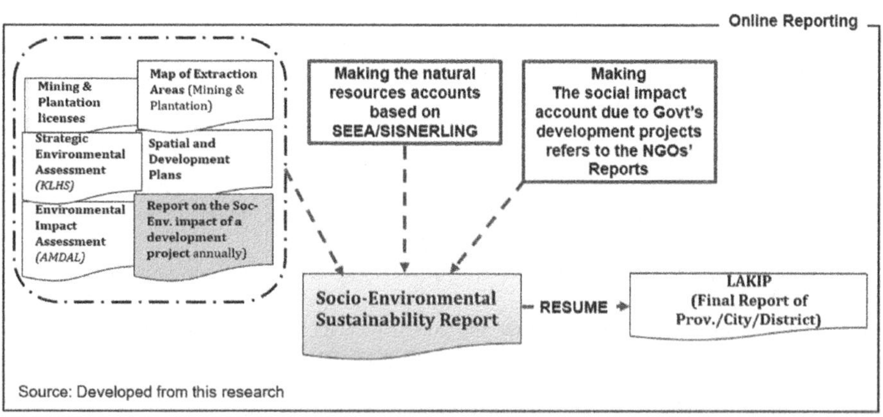

Fig. 6.4 Framework of Local Government's Socio-Environmental Sustainability Report

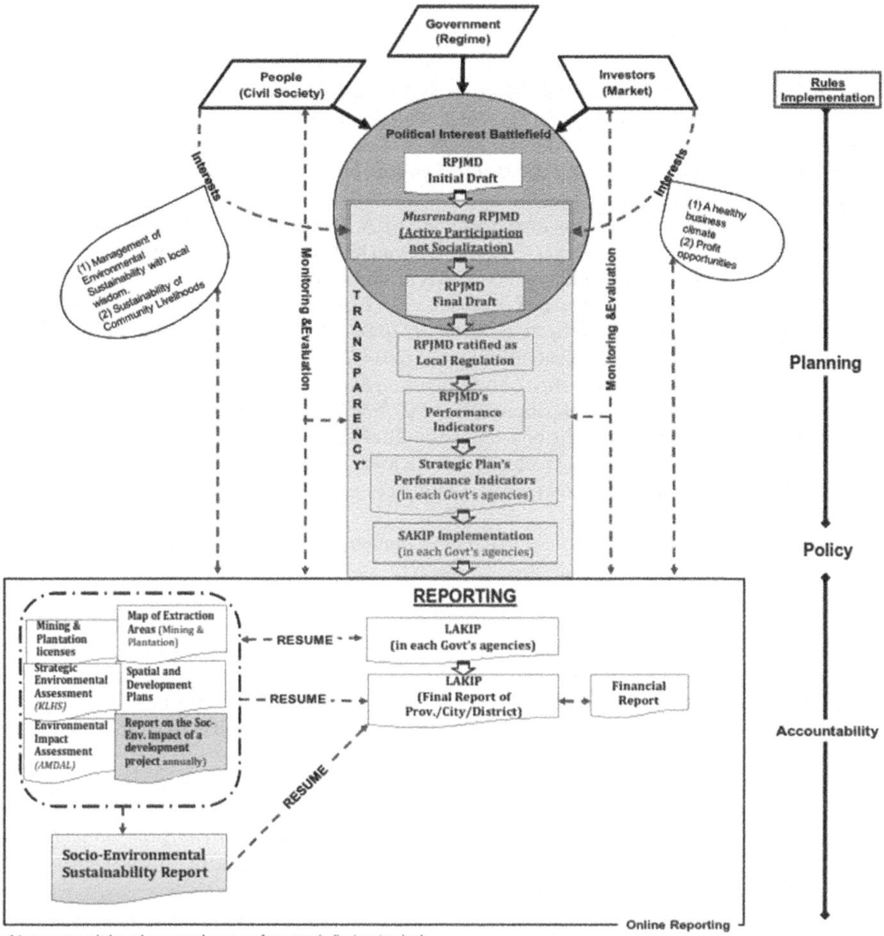

* transparency is based on agreed-upon performance indicator standards between governments and elements of society

Source: Developed from this research

Fig. 6.5 Towards Local Government's Integrated Accountability Framework

Closing this chapter, I would like to quote views of two Nobel laureates, namely Prof. Joseph E. Stiglitz and Prof. Amartya Sen, along with Prof. Jean-Paul Fitoussi who is a member of the French Commission for Measurement of Economic Performance and Social Progress (2008–2009), who also has a great concentration on presenting sustainable social-environmental measures. They argued:

> However, whatever we do, the dashboard and the index are only one part of the story. Most of the efforts to study sustainability focus on increasing our knowledge of how the economy and the environment interact now and in the future (Stiglitz et al., 2009).

References

Aspinall, E. (2013). A NATION IN FRAGMENTS: Patronage and Neoliberalism in Contemporary Indonesia. *Critical Asian Studies, 45*(1), 27–54. https://doi.org/10.1080/14672715.2013.758820.

Bonsón, E., Torres, L., Royo, S., & Flores, F. (2012). Local e-government 2.0: Social media and corporate transparency in municipalities. *Government Information Quarterly, 29*(2), 123–132. https://doi.org/10.1016/j.giq.2011.10.001.

BPS. (2017). Indonesia experience in implementing SEEA. *Seminar on "The Scaling up of SEEA—Implementation towards 2020 Targets"*. Retrieved from https://unstats.un.org/unsd/statcom/48th-session/side-events/documents/20170306-2L-Kecuk-Suhariyanto.pdf.

BPS. (2018). *Sistem Terintegrasi Neraca Lingkungan dan Ekonomi Indonesia 2013–2017*. Retrieved from https://www.bps.go.id/publication/2019/12/27/53e7360b49c122c58475d2c9/sistem-terintegrasi-neraca-lingkungan-dan-ekonomi-indonesia-2014-2018.html.

BPS. (2019). *Sistem Terintegrasi Neraca Lingkungan dan Ekonomi Indonesia 2014–2018*. Retrieved from https://www.bps.go.id/publication/2019/12/27/53e7360b49c122c58475d2c9/sistem-terintegrasi-neraca-lingkungan-dan-ekonomi-indonesia-2014-2018.html.

Esser, F. (2013). Mediatization as a challenge: Media logic versus political logic. In *Democracy in the age of globalization and mediatization. Challenges to democracy in the 21st century series*. London: Palgrave Macmillan.

Evans, P., & Stephens, J. D. (1988). Studying development since the sixties. *Theory and Society, 17*(5), 713–745.

Giddens, A. (2013). *The third way: The renewal of social democracy*. Wiley.

Harun, H., An, Y., & Kahar, A. (2013). Implementation and challenges of introducing NPM and accrual accounting in Indonesian local government. *Public Money and Management, 33*(5), 383–388. https://doi.org/10.1080/09540962.2013.817131.

Islam, M., & Deegan, C. (2010). Media pressure and corporate disclosure of social responsibility performance: A case study of two global clothing and sports retail companies. *Accounting Business Research, 40*(2), 131–148.

KPA. (2016). *Catatan Akhir Tahun 2016: Liberalisasi agraria diperhebat, reforma agraria dibelokkan*. https://doi.org/10.1017/CBO9781107415324.004.

KPA. (2018). *Catatan Akhir Tahun 2018: Masa Depan Reforma Agraria Melampaui Tahun Politik*. Retrieved from https://kpa.or.id/publikasi/baca/laporan/30/Catahu_2018:_Masa_Depan_Reforma_Agraria_Melampaui_Tahun_Politik/?lang=id.

KPA. (2019). *Laporan Ringkas: Empat Tahun Implementasi Reforma Agraria*. Retrieved from https://kpa.or.id/publikasi/baca/laporan/77/Empat_Tahun_Implementasi_Reforma_Agraria/?lang=id.

Lanis, R., & Richardson, G. (2013). Corporate social responsibility and tax aggresiveness: A test of legitimacy theory. *Accounting, Auditing & Accountability Journal, 26*(1), 75–100.

LBH-Jakarta. (2015). *Atas Nama Pembangunan: Laporan Penggusuran Paksa di Jakarta Tahun 2015* (Vol. 2017). Retrieved from https://www.bantuanhukum.or.id/web/atas-nama-pembangunan-laporan-penggusuran-paksa-di-jakarta-tahun-201.

LBH-Jakarta. (2017). *Seperti Puing: Laporan Penggusuran Paksa di Wilayah DKI Jakarta Tahun 2016*. Retrieved from https://www.bantuanhukum.or.id/web/wp-content/uploads/2017/04/Laporan-Penggusuran_LBHJAKARTA_2016.pdf.

Malaka, T. (1921). *Parlemen atau Soviet* (2014, Ed.). Jakarta: Sega Arsy.

Malaka, T. (1926). *Aksi Massa*. Retrieved from https://www.marxists.org/indonesia/archive/malaka/AksiMassa/.

Malaka, T. (1943). *Madilog*.

Misztal, B. A. (2001). Trust and cooperation: The democratic public sphere. *Journal of Sociology, 37*(4), 371–386. https://doi.org/10.1177/144078301128756409.

Pina, V. (2005). E-government and the transformation of public administrations in EU countries: Beyond NPM or just a second wave of reforms? *Online Information Review, 29*(5), 531–553.

Pina, V., Torres, L., & Royo, S. (2010). Is E-Government leading to more accountable and transparent local governments? An overall view. *Financial Accountability & Management, 26*(1), 3–20. https://doi.org/10.1111/j.1468-0408.2009.00488.x.

Sanusi, A., Butler, B., & Rola-rubzen, M. F. (2013). Local politics and business in Indonesia: A non-market strategy approach. *Provincial governance strengthening programme supported by UNDP Indonesia* (pp. 1–20). Unika Atmajaya, Jakarta: Unika Atmajaya.

Shahib, H. M., & Risky, F. R. (2017). Accountability in the internet Era: A lesson from local governments in Indonesia. *Hasanuddin Economics and Business Review, 1*(1), 57–74.

Stiglitz, J. E., Sen, A., & Fitossi, J.-P. (2009). *Commission on the measurement of economic performance and social progress report-2009.* Paris: The Commission on The Measurement of Economic Performance and Social Progress-France.

United Nations. (2014). *System of environmental-economic accounting 2012: Central framework.* New York: United Nations.

WALHI. (2015). *Tinjauan Lingkungan Hidup 2015 WALHI*. Retrieved from https://walhi.or.id/wp-content/uploads/2016/12/OutLook2015_W.pdf.

WALHI. (2016). *Tinjauan Lingkungan Hidup WALHI 2016: Keharusan Pembenahan Struktural untuk Perbaikan Tata Kelola.* Retrieved from Wahana Lingkungan Hidup Indonesia website: https://www.walhi.or.id/2016/12/22/outlook-2016-tujuan-lingkungan-hidup/.

WALHI. (2018). *Tinjauan Lingkungan Hidup 2018 : Bagaimana Masa Depan Keadilan Ekologis 2018?* Retrieved from https://walhi.or.id/wp-content/uploads/2018/12/Layout_Tinjauan-Lingkungan-2018.pdf.

Walhisulsel. (2019). Aliansi Selamatkan Pesisir (ASP) Tolak Tambang Pasir Laut, Revisi RZWP3K Sulsel! Retrieved August 1, 2019, from https://walhisulsel.or.id/2533-press-rilis-aliansi-selamatkan-pesisir-asp-tolak-tambang-pasir-laut-revisi-rzwp3k-sul-sel/.

Chapter 7
Closing Remarks

Please keep the earth and all of its contents:
The sky, humans, and forests.
—Kajang Old Proverb

7.1 Conclusions

Finally,[1] this chapter will explain the conclusions of the main findings as well as the synthesis results of this research to develop a new integrated accountability framework in the local governments that better accommodates aspirations of the social-environmental NGO advocacy movement and the people affected by social-environmental issues. That is in line with the spirit of the emancipatory socio-environmental accounting idea.

This study found that the current local government accountability system in Indonesia is highly dominated by the logic of NPM which is rooted strongly in Neoliberalism. The characteristics of NPM logic hegemony, which are highly market-oriented and similar to business-oriented entities that are highly focused on achieving efficient-effective measures based on monetary values in managing their resources, are found starting from regulations, practices of determining the strategic objectives of local governments, policies, to the practice of accountability of local government performance. This hegemony of NPM logic, that makes the aspirations of people affected by social-environmental problems along with NGOs that are actively advocate for these issues, tend not to have a place in the government performance accountability system that is integrated from planning to reporting the accountability reports of local government performance.

[1] Kajang is one of the indigenous people in Bulukumba district, South Sulawesi Province, Indonesia. These people are well known as guardians of forest in Bulukumba whose rich of local wisdom in managing their environment.

H. M. Shahib, *Towards the Local Government's Integrated Accountability Framework*,
https://doi.org/10.1007/978-981-33-6617-6_7

Thus, in order to achieve an emancipatory-integrated accountability framework in local government as an antithesis of NPM hegemony, there are three components must be considered and applied. They are:

First, it is essential to present the substantial participation and transparency, as the basis to gather the aspirations of affected people and socio-environmental NGOs, to be the reference from planning, practices, to accountability report. Especially, there is a strong allegation about the presence of ruling regime's political interest influence in the government development plan (RPJMD), which then leads to the sense that development activities and government accountability practices do not accommodate people's aspirations.

Furthermore, these following points are the summary of the main values strived for by NGOs and communities affected by social-environmental issues as found by this research: (1) they encourage the presence of government environmental management that gives space to the local wisdom of the community and (2) ensures the sustainability of community livelihoods resources; they encourage (3) transparency and (4) in fair community participation with the government from planning to implementation of development in each region; and (5) they encourage the presence of law enforcement, especially environmental law for the achievement of ecological justice in people's lives.

Second point, this research believes that the optimization of data facilities and information technology owned by the government can assist changes in the government's performance accountability system towards a more integrated and emancipating public aspirations. This is because, according to the findings of this study, regulations and instruments that support equitable and sustainable environmental management are basically already available. For example, maps of extraction areas (mining and plantation), spatial development plans, mining and plantation licenses list, strategic environmental assessment *(KLHS)*, environmental impact assessment *(AMDAL)*, and government's electronic data systems are available.

Unfortunately, the government political will has so far inclined to be oriented towards economic growth alone, not yet directed towards utilization of these instruments for the sake of achieving ecologically, equitable, and emancipative development. This situation is proofed by the difficulty of the community to access information related to environmental management, specifically in regions with social-environmental issues. This study found the fact that transparency and accountability of the government on social-environmental issues are yet represent the aspirations of affected elements of society. Thus, the government must optimize information technology, especially the use of internet to improve the quality of public participation and government transparency in planning, activities and reporting government accountability.

Third, as the final point that must be considered and implemented, this study tries to provide the input by optimization of initial information about the environmental management hold by the local government. The government should be able to present a report on the socio-environmental impact of a development project (periodically) to explain its impact to the affected people in particular, and the public in general. Also, the government should be able to present social-environmental sustainability reports (periodically) with instruments already owned by the government. This is important in order to measure the extent of impact of government development projects together with the private sector on the environmental and social security of people. The government can develop it based on the internationally accepted SEEA (the statistical framework of natural resource balance). And for its social impact, the government can adopt Indonesia's leading social-environmental NGOs reporting models. This is important to increase public engagement to government activities. At the same time, it will maintain the legitimacy of the government in the public eyes.

7.2 Research Implications

This research provides several implications to the development of theory-knowledge, policy, and practice of government. Specifically, related to the development of local government accountability discourse on the issue of sustainable development. Here are important implications of this research:

1. *Theoretical and Philosophical Implications*
   This research provides several important notes related to the development of social-environmental accountability discourse in the accounting and accountability literature. First, this study looks at both the central and regional governments as the most responsible actor for the issue of sustainability, as the government has the authority over planning, licensing, supervision, and law enforcement on social-environmental issues. On that basis, it becomes important to present a social-environmental accountability model that can be used as a guide by the people to assess the government's social-environmental performance comprehensively.
   Second, this research finds the fact that the strength and active political movement of all government stakeholders largely determines the direction of the government's social-environmental accountability model development. Therefore, the discussion of social-environmental accountability discourse can no longer only by looking at the appearance of the accountability model alone or discuss it with the lens of theory that denies the active political movements of each stakeholder.
   Third, the consequences of having so much political power influence in determining the government's social-environmental accountability model leads to the need to bring substantial participation and transparency from the planning process to the practice of accountability, as it is seen to be able to avoid the presence of information gaps in any stakeholders. Do not forget, information is the main weapon to win the battle of political discourse. This ancient idiom will continue to have its place "they who rule the media (information) will rule the world".
   Finally, all these findings above fill the gap in the area of emancipatory socio-environmental accounting framework discourse. It is related to the studies of local government social-environmental accountability system that focus in managing

the wealth of natural resources and vulnerability of social-environmental problems.
From a philosophical point of view, this research contributes to introduce an Indonesian-critical paradigm, from the critical thinking of Tan Malaka, a philosopher as well as a hero of Indonesia's independence. This critical thinking of Tan Malaka rests on (1) the view of the importance of making criticisms and adjustments to all foreign concepts that are going to be applied in Indonesian government practices to ensure they are in accordance with the values and interests of the Indonesian people. (2) It also emphasizes the importance of the aspirations of all classes in society (especially small communities) such as farmers, fishermen, laborers, and the poor, to be present in the structure of government management system. Finally, (3) Tan Malaka's thoughts emphasize the importance to understand the active political movements between social structures (government systems) and agents (individual/society) in shaping a social order, such as regulations and policies which will ultimately affect the lives of many people.
As far as my knowledge, this study is the first one to develop and use the critical methodology of Tan Malaka as a research analysis tool in the fields of accounting and accountability. I hope, in the future, Tan Malaka's critical methodology will be widely used as an analytical tool by researchers in accounting, management, and governance in Indonesia in particular, and in the world at large.

2. *Implication to policies*
   In the realm of policy, this research is expected to contribute to Indonesian local government institutions as a new knowledge base to develop integrated accountability policies that accommodates aspirations of people affected by social-environmental problems resulting from local government development policies. Second, in terms of the central government, especially to regulatory and supervisory institutions for the implementation of the local government performance accountability system (KEMENPAN-RB), this research can become one of the knowledge bases to develop policies and regulations by them for the perfection of government performance accountability systems in the future, that accommodates more aspirations of Indonesian people. In addition, as the social-environmental conflict is the problem of all humanity these days, this research is also expected to be able to inspire other countries, especially countries relying on exploration of natural resources and loaded with natural resources conflicts, to improve their social-environmental accountability policies.
3. *Implication to Praxis*
   In line with the implications to policy, the praxis implication of this research is expected to become a new guidance to develop the government's integrated accountability system, particularly as a way of new approach, which better accommodate people's aspirations on the development plans to present a report in the scope of local government accountability. On the other hand, this research is expected to be the basis of new perspectives related to how the community and government should interact to reach a consensus related to accountability for social-environmental issues in Indonesia. In addition to that, it is hoped that the results of this study can be used as a global campaign tool for sustainable

development that is aligned with people's interests, based on their respective social-environmental issues.

7.3 Limitations and Recommendations for Future Research

This study has limitations that are in line with the choice of research methodology used in this study. First, there were several obstacles to get all study locations as previously proposed. For example, there was no feedback in communications with several activists and local people in several locations, and a great earthquake in Palu and Donggala (Central Sulawesi) in September 2018 created severe damage and difficulties to access these locations. Also, I faced limited time and funding. As the consequence of these obstacles, this study was only able to get into the coastal reclamation in Makassar (south Sulawesi), as one main location to be studied deeply in relation with the socio-environmental activists' advocacy activities, government's accountability practices, and the affected people. Furthermore, for other study objects, which cannot be reached, this research tries to discuss them by using other sources such as the annual report of socio-environmental NGOs, news, and government's accountability documents related to those socio-environmental issues.

Second, this research is limited by the observation period and the time period of the research sources (the condition of the observed area, documents, interviews, and reports). Thus, it is possible if there are changes on the sources of this research in the future. Specifically, for interview to the local government's officers, in this study, I found several government officers that refused me to interview them and not too welcome in giving the real answers. This situation then made this study tend to focus on analyzing the government's documents related to the research's issues.

From the research limitations, I then propose several recommendations for further research. First, the next study should prepare the research schedule as effective as possible, specifically to ensure the object and informant of the research and not to forget to set the Plan-B approach in order to keep the focus on the study objectives. Second, as this research focuses on criticizing the Indonesian local governments' framework and system of accountability, future research can focus more on things in more micro-scope, such as development of social-environmental index measurements that can be applied in government accountability reports. Also, the future study can focus on comparing the other government's frameworks of accountability between two states that have similiarity from its political and socio-environmental issues to gather the wide understanding. Third, to overcome the limitation of observation period and the time period of the research sources, it may be good to try to consider using quantitative or mix-method approach. Last, the next research should be prepared to find a good interpersonal approach to the informant in order to gain substantive information, as sometimes they are reluctant to give such information.

Finally, to close this chapter, I would like to quote Tan Malaka's words "The way to get results is more important than the results themselves" (Malaka, 1943:99). Hopefully this research can become a catalyzer for other significant achievements in the discourse of social-environmental accountability in Indonesia in particular, and in the world in general.

Reference

Malaka, T. (1943). *Madilog*.

The manufacturer's authorised representative in the EU is Springer Nature Customer Service Centre GmbH, Europaplatz 3, 69115 Heidelberg, Germany. If you have any concerns regarding our products, please contact ProductSafety@springernature.com

Printed and bound by CPI Group (UK) Ltd, Croydon, CR0 4YY
30/09/2026
02225306-0009